THE FLOCK

Western Literature Series

DON JOSÉ'S DRIVE

"All through the dark they steered a course by the stars"

THE FLOCK

BY
MARY AUSTIN

ILLUSTRATED BY E. BOYD SMITH

With an Afterword by Barney Nelson

UNIVERSITY OF NEVADA PRESS
RENO & LAS VEGAS

Western Literature Series

The Flock was first published by Houghton, Mifflin and
Company in the United States in 1906.
University of Nevada Press, Reno, Nevada 89557 USA
www.unpress.nevada.edu
New material copyright
© 2001 by University of Nevada Press
Illustrations in the new material copyright
© 2001 by The Huntington Library

Manufactured in the United States of America
Cover design by Carrie House

CIP data appear at the end of the book

The paper used in this book meets the requirements of
American National Standard for Information Sciences—
Permanence of Paper for Printed Library Materials,
ANSI Z39.48-1984. Binding materials were selected for
strength and durability.

University of Nevada Press Paperback Edition, 2001

This book has been reproduced as a digital reprint.

Frontispiece: Don José's Drive, by E. Boyd Smith

DEDICATED TO
THE FRIENDLY FOLK IN INYO
AND
THE PEOPLE OF THE BOOK

CONTENTS

I

THE COMING OF THE FLOCKS — HOW RIVERA Y MONCADA BROUGHT THE FIRST OF THEM TO ALTA CALIFORNIA, AND A PREFACE WHICH IS NOT ON ANY ACCOUNT TO BE OMITTED.

CHAPTER I

THE COMING OF THE FLOCKS

A GREAT many interesting things happened about the time Rivera y Moncada brought up the first of the flocks from Velicatá. That same year Daniel Boone, lacking bread and salt and friends, heard with prophetic rapture the swaying of young rivers in the Dark and Bloody Ground; that year British soldiers shot down men in the streets of Boston for beginning to be proud to call themselves Americans and think accordingly; that year Junípero Serra lifted the cross by a full creek in the Port of

Monterey; — coughing of guns by the eastern sea, by the sea in the west the tinkle of altar bells and soft blether of the flocks.

All the years since Oñate saw its purple hills low like a cloud in the west, since Cabrillo drifted past the tranquil reaches of its coast, the land lay unspoiled, inviolate. Then God stirred up His Majesty of Spain to attempt the dominion of Alta California by the hand of the Franciscans. This sally of the grey brothers was like the return of Ezra to upbuild Jerusalem; "they strengthened their hands with vessels of silver," with bells, with vestments and altar cloths, with seed corn and beasts collected from the missions of Baja California. This was done under authority by Rivera y Moncada. "And," says the Padre in his journal, "although it was with a somewhat heavy hand, it was undergone for God and the King."

Four expeditions, two by land and two by sea, set out from Old Mexico. Señor San José being much in the public mind at that time, on account of having just delivered San José del Cabo from a plague of locusts, was chosen patron of the adventure, and Serra, at the re-

quest of his majesty, sang the Mass of Supplica·
tion. The four expeditions drew together again
at San Diego, having suffered much, the ships'
crews from scurvy and the land parties from
thirst and desertion. It was now July, and back a
mile from the weltering bay the bloom of cacti
pricked the hot, close air like points of flame.

Señor San José, it appeared, had done enough
for that turn, for though Serra, without waiting
for the formal founding of Mission San Diego
de Alcalá, dispatched Crespi and Portolá northward, their eyes were holden, and they found
nothing to their minds resembling the much
desired Port of Monterey, and the Mission
prospered so indifferently that their return was
to meet the question of abandonment. The
good Junípero, having reached the end of his
own devising, determined to leave something
to God's occasions, and instituted a *novena*.
For nine days Saint Joseph was entreated by
prayers, by incense, and candle smoke; and
on the last hour of the last day, which was
March 19, 1770, there appeared in the far blue
ring of the horizon the white flick of a sail
bringing succor. Upon this Serra went on the

second and successful expedition to Monterey, and meantime Don Fernando de Rivera y Moncada had gone south with twenty soldiers to bring up the flocks from Velicatá.

Over the mesa from the town, color of poppies ran like creeping fire in the chamisal, all the air was reeking sweet with violets, yellow and paling at the edges like the bleached, fair hair of children who play much about the beaches. Don Fernando left Velicatá in May — O, the good land that holds the record of all he saw! — the tall, white, odorous Candles - of - Our

Lord, the long, plumed reaches of the chamisal, the tangle of the *meghariza*, the yellow-starred plats of the *chili-cojote*, reddening berries of rhus from which the Padres were yet to gather wax that God's altars might not lack candles, the steep barrancas clothed with deer-weed and *toyon*, blue hills that swam at noon in waters of mirage. There was little enough water of any sort on that journey, none too much of sapless feed. Dry camp succeeded to dry camp. Hills neared them with the hope of springs and passed bone-dry, inhospitably stiff with cactus and rattle weed. The expedition drifted steadily northward and smelled the freshness of the sea; then they heard the night-singing mocking bird, wildly sweet in the waxberry bush, and, still two days from San Diego, met the messengers of Governor Portolá going south with news of the founding of Monterey. This was in June of 1770. No doubt they at San Diego were glad when they heard the roll of the bells and the blether of the flock.

Under the Padres' careful shepherding the sheep increased until, at the time of the secularization, three hundred and twenty thousand fed in the Mission purlieus. Blankets were woven, serapes, and a coarse kind of cloth called *jerga*, but the wool was poor and thin; probably the home government wished not to encourage a rival to the exports of Spain. After secularization in 1833, the numbers of sheep fell off in California, until, to supply the demand for their coarse-flavored mutton, flocks were driven in from Mexico. These "mustang sheep" were little and lean and mostly black, sheared but two and one half pounds of wool, and were so wild that they must be herded on horseback. About this time rams were imported from China without materially improving the breed. Then the rush westward in the eager fifties brought men whose trade had been about sheep. Those who had wintered flocks on New England hill pastures began to see possibilities in the belly-deep grasses of the coast ranges. In '53, William W. Hollister brought three hundred ewes over the emigrant trail and laid the foundation of a fortune. But think of the fatigues of it, the rivers to swim, the passes to attempt, the watch fires, the far divided water holes, the interminable lapsing of days and nights, — and a sheep's day's journey is

seven miles! No doubt they had some pressing, and comfortable waits in fat pastures, but it stands on the mere evidence of the fact, that Hollister was a man of large patience. During the next year Solomon Jewett, the elder, shipped a flock by way of Panama, and the improvement of the breeds began. The business throve from the first; there are men yet to tell you they have paid as high as twelve dollars for a well-fatted mutton.

The best days of shepherding in California were before the Frenchmen began to appear on the mesas. Owners then had, by occupancy, the rights to certain range, rights respected by their neighbors. Then suddenly the land was overrun by little dark men who fed where feed was, kept to their own kind, turned money quickly, and went back to France to spend it. At evening the solitary homesteader saw with dread their dust blurs on his horizon, and at morning looked with rage on the cropped lands that else should have nourished his own necessary stock; smoke of the burning forests witnessed to heaven against them. Of this you shall hear further with some particularity.

Those who can suck no other comfort from the tariff revision of the early eighties may write to its account that it saved us unmeasured acreage of wild grass and trees.

What more it did is set down in the proper place, but certainly the drop in prices drove out of the wool industry those who could best be spared from it. Now it could be followed profitably by none but the foreseeing and considering shepherd, and to such a one dawned the necessity of conserving the feed, though he had not arrived altruistically at wanting it conserved for anybody else. So by the time sheep-herding had recovered its status as a business, the warrings and evasions began again over the withdrawal of the forest reserves from public pasturing. Here in fact it rests, for though there be sheep-owners who understand the value of tree-covered water-sheds, there are others to whom the unfair discrimination between flocks and horned cattle is an excuse for violation; and just as a few Cotswolds can demoralize a bunch of tractable merinos, so the unthinking herder brings the business to disesteem.

What I have to do here is to set down without prejudice, but not without sympathy, as much as I have been able to understand of the whole matter kindled by the journey up from Velicatá in the unregarded spring of 1770, and now laid to the successors of Don Fernando de Rivera y Moncada.

I suppose of all the people who are concerned with the making of a true book, the one who puts it to the pen has the least to do with it. This is the book of Jimmy Rosemeyre and José Jesús Lopez, of Little Pete, who is not to be confounded with the Petit Pete who loved an antelope in the Ceriso, — the book of Noriega, of Sanger and the Manxman and Narcisse Duplin, and many others who, wittingly or unwittingly, have contributed to the performances set down in it. Very little, not even the virtue of being uniformly grateful to the little gods who have constrained me to be of the audience, can be put to the writer's credit. All of the book that is mine is the temper of mind which makes it impossible that there should be any play not worth the candle.

By two years of homesteading on the borders of Tejon, by fifteen beside the Long Trail where it spindles out through Inyo, by all the errands of necessity and desire that made me to know its moods and the calendar of its shrubs and skies, by the chances of Sierra holidays where there were always bells jangling behind us in the pines or flocks blethering before us in the meadows, by the riot of shearings, by the faint winy smell in the streets of certain of the towns of the San Joaquin that apprises of the yearly inturning of the wandering shepherds, I grew aware of all that you read here and of much beside. For if I have not told all of the story of Narcisse Duplin and what happened to the Indian who worked for Joe Espelier, it is because it concerned them merely as men and would as likely have befallen them in any other business.

Something also I had from the Walking Woman, when that most wise and insane creature used to come through by Temblor, and a little from pretty Edie Julien interpreting shyly in her father's house, but not much, I being occupied in acquiring a distaste for my

own language hearing her rippling French
snag upon such words as " spud " and " bunch "
and "grub." In time I grew to know the owner
of flocks bearing the brand of the Three Legs
of Man, and as I sat by his fire, touching his
tempered spirit as one half draws and drops a
sword in its scabbard for pleasure of its fine-
ness, becoming flock-wise I understood why
the French herders hereabout give him the
name of the Best Shepherd. I met and talked
with the elder Beale after he had come to the
time of life when talking seems a sufficient
occupation, and while yet there was color and
glow as of the heart wood breaking in the
white ash of remembrance. But, in fact, the
best way of knowing about shepherding is to
know sheep, and for this there was never an
occasion lacking. In this land of such indolent
lapping of the nights and days that neither
the clock nor the calendar has any pertinence
to time, I call on the eye of my mind, as it
were, for relief, looking out across the long
moon-colored sands, and say : —

" Do you see anything coming, Sister Anne ? "
" I see the dust of a flock on the highway."

Well, then, if from the clutch of great Te-
dium (of whom more than his beard is blue)
there is no rescue but such as comes by way
of the flock, let us at least miss no point of the
entertainment.

II

 THE SUN IN ARIES — WHICH RELATES HOW THE FLOCKS COME TO THE HOME PASTURES, AND THE PROPER MANAGEMENT OF LAMBS.

CHAPTER II

THE SUN IN ARIES

ABOUT the time there begin to be cloud
shadows moving on the unfurrowed wild pastures of the San Joaquin there begin to be
windless clouds of dust coasting the foothills
under the Sierras, drifting in from the blue
barriers of the seaward ranges, or emerging
mysteriously from unguessed quarters of the
shut horizon. They drop into the valley from
Tehachapi, from Kings River and Kern, as
far driven as from the meadows of Mono and

Yosemite, and for the time of their coming
acknowledge no calendar but the unheralded
Beginning of Rains. Let there be but the
faintest flush of green on the pastures they
left bare in the spring, and by some wireless
prescience all the defiles of Little Lake and
Red Rock are choked with the returning
flocks. Let one of the pallid fogs of early winter obscure the hollow of the valley for a night
and a day, and at its clearing, mark the un-
patented lands all freckled with dust-colored
bands. Drenched mornings one counts a
dozen pale blurs of moving dust low along the
foothills, and evenings on the red track of the
sun sees the same number of shepherd fires
blossom through the dusk. The count of them
diminishes yearly, but since as long ago as the
early sixties, the southern end of the San Joaquin Valley has been the favorite lambing
place of flocks ranging north and east as many
miles as a flock can cover in the nine or ten
months' interval between the end and beginning of winter feed. The equable weather, the
great acreage of unclaimed pasture, and the
nearness of the trains that pound through the

valley like some great, laboring, arterial beat
of the outer world, draw the wandering flocks
to a focus once in the year about the time the
sun enters Aries. As I say, they acknowledge
no calendar but the rains, and the earlier these
come the better, so that the flocks get into the
home pastures before the ewes are too heavy
for traveling. Before all, at lambing time the
shepherd seeks quiet and good pasture, and if
he owns no land at all he must at least have
a leasehold on suitable places to put up his
corrals.

Since as long ago as men referred their affairs to the stars February has been the month
for lambing, and that, you understand, is as long
ago as the sun was actually in Aries, before the
precession of the Equinoxes pulled it back
along the starry way. At Los Alisos the middle of January sees the ewes all gathered to the
home ranch, and here and there from deep
coves of the hills, yellowing films of dust rising
steadily mark where the wethers still feed, fattening for the market. At this time of the year
the land is quiescent and the sky clearer than
it will be until this time again, halting midway

between the early rains and late. All the summer's haze lies folded in a band a little above the
foothills and below the snows of the Sierras, so
that the flame-white crests appear supernatu-
rally suspended in clearness, the very front and
battlements of heaven. In the fields above the
little green tumuli of alfalfa, great cottonwoods
click a withered leaf or two, and the tops of
the long row of close, ascending poplars, running down from the ranch house, are absorbed
in an infinite extension of light. Now besides
the weirs one finds a heron's feather, and mal-
lards squatter in the crescent pools below the
drops. The foothills show greenness deepening in the gullies; nights have a touch of chill-
iness with frequent heavy dews.

Leberge, the head shepherd of Los Alisos, is
a careful man. The ewes from which lambs are
first expected have the fattest pastures; corrals
to accommodate a hundred of them are set off
with movable fencing; the number of herders
is multiplied and provided with tar and tur-
pentine and such remedial simples. But for
the most part nature has a full measure of
trust. In the north where sheep run on fenced

pastures, the mothers have leave to seek shelters of rock and scrub and clear little formless hollows to bed their young. There shepherding has not wholly superseded the weather wisdom of the brute, and in years of little promise the untended ewes will not lick their lambs. But here among the hobo herds of the Long Trail, artificial considerations, such as the relative price of wool and mutton and the probable management of forest reserves, determine whether the ewe shall be allowed to rear the twin lambs that nature allots her. Years of curtailed pastures she cannot suckle both and grow wool, and neither youngster will be strong enough to endure the stress of a dry season : the mother becomes enfeebled, and the too grasping shepherd may end by losing all three. Much depends on the promptness with which the weaker of twins is discarded or suckled to some unfortunate mother of stillborn lambs. Once a ewe has smelled the smell of her offspring the herder must take a leaf out of the book of the Supplanter in the management of forced adoptions. The skin of the dead lamb is sewed about the body of the foundling, limp little legs dangling about its legs, a stiff little tail above a wagging one, — all of no moment so long as the ewe finds some rag-tag smell of her own young among the commingling smells of the stranger and the

dry and decaying hide.

Here and there will be young ewes in their first season refusing their lambs. Trust the French herders for finding devices against such a reversion of nature. About the corners of the field will be pits where by enforced companionship the one smell of all smells a sheep must remember, with no root in experience or memory, gropes to the seat of her dull consciousness, and the ewe gives down her milk. A commoner device is to tie the recalcitrant dam near a dog, and the silly sheep, trembling and afraid, too long a mere fraction of a flock to have any faculty for sustaining dread, makes friends with her unwelcome lamb as against their common enemy, the collie. Remedial measures such as these must be immediate, otherwise in chill nights of frost or weeping fog, the unlicked, unsuckled lambs will die. So it is that here and there, but not invariably, one sees a shepherd making rounds with a lantern through the night, and in a flock of three to five hundred ewes finding much to do.

Nights such as this the bunch grass cowers to the wind that lies too low along the pasture to stir the tops of trees. The Dipper swings low from the Pole, and changeful Algol is a beacon in the clear space between the ranges above which the white planets blink and peer. The quavering mu-uh-uh, mu-uh-uh-uh of the mothering ewes keeps on softly all night. The red eye of the herder's fire winks in the ash; the dogs get up from before it, courting an invitation to their accustomed work. Whining throatily, they nose at the master's heels and are bidden down again lest they scare the ewe from her unlicked lamb. Great Orion slopes from his meridian, and Rigel calls Aldebaran up the sky. The lantern swings through the dark sweep of pasture, cool and dewy and palpitant with the sense of this earliest, elemental stress of parturition.

Every now and then some unconsidered protest arises against the clipped and mutilated speech by which a human mother expresses her sense of satisfaction in her young. But let the protestant go to Los Alisos when the sun is far gone in its course in Aries, and understand, if he can, the breaking of the sheep's accustomed bleat to the soft mutter of the ewes, and what over-sense prompts the wethers to futile adoptions of lambs coaxed from the dam by the same soft, shuddering cry. Such a sheep is by herders called a "grannie," and by simply saying it is so, passed by, but at this hour when the darkness is impregnate with the dawn and the sense responds to the roll of the world eastward, the return of these unsexed brutes to the instinct of parental use takes on the proportions of immeasurable law. But nourishing is in fact the greater part of mothering, and lest it should come amiss the herder marks the careless or unwilling ewe and the lamb each with a black daub on the head or shoulder, pair and pair alike, and conspicuously, so that he sees at a glance at nursing time that each young goes to its own dam.

Young lambs are principally legs, the connecting body being merely a contrivance for converting milk into more leg, so you understand how it is that they will follow in two days and are able to take the trail in a fortnight, traveling four and five miles a day, falling asleep on their feet, and tottering forward in the way. By this time it has become necessary to move out from the home fold to fresher pastures, but keeping as close as the feed allows. Not until after shearing do they take to the mountain pastures and the Long Trail. Now there will be bird's-eye gilias, sun-cups, and miles of pepper grass on the mesas; coastward great clots and splashes of gold, glowing and dimming as the sun wakes the dormidera or the mist of cloud folds it up. Wethers and yearlings will be ranging all abroad, but ewes with lambs, five or six hundred in a bunch, will be kept as much as possible in fenced pastures. At a month old the flock instinct begins to stir; lambs will run together and choose a bedding place sunward of a fence or the windbreak of young willows along an irrigating ditch. Here they leap and play and between

whiles doze. Here the ewes seek them with
dripping and distended udders. It is a question during the first week if the lamb knows
its mother at all and she it by smell only, and
smells indiscriminately at black lambs or white,
but at the end of eight days they come calling
each to each. Let three or four hundred lambs
lie adoze in the sun of a late afternoon; comes
a ewe across the pastures, craving relief for
her overflowing dugs. Yards away the lamb
answers her out of sleep and goes teetering
forward on its rickety legs, her own lamb, mind
you, capering up with perhaps the tattered
skin askew on its back, that first deceived her
into permitting its hungry mouth; and not
one of the four hundred others has more than
flicked an ear or drawn a deeper breath. But
suppose her to have twins, these will have

been tied together by the herder so that the stronger may not get first to the fountain but drags his weaker brother up. In time the conviction of two mouths at the udder becomes rooted, and one will not be permitted without the other. Then the amount of urgency to come on and be fed which the spraddle-kneed first comer can put into the waggings of his tail, hardly bears out the observation that the twins do not know each other very well except by smell.

The Valley of the San Joaquin is wide enough to give the whole effect of unmeasured plain, and the sky at the end of the lambing season shallow, and hemmed by tenuous cloud. Close-shut days the flocks drift about its undulations, sandy, shelterless stretches, dull rivers defiled by far-off rains, one day east under black, broad-heading oaks, another west in foolish, oozy intricacies of sloughs where rustling tules lean a thousand ways. Blossoms come up and the lambs nibble them; filaree uncurls for the sheep to crop. The herder walks at the head of the flock, and if he is near enough, watches the hilltops breaking the thin woof of cloud to note how the feed advances in their deepening green; and always he prays for rain. At intervals the head shepherd bears down upon him by some of the whity-brown roads that run every way in the valley and by endless crisscrossing and ramifications lead to all the places where you do not particularly wish to go. Now and then a buyer reaches him by the same roads to overlook the yearlings or estimate the chances of wool. Rains may come as late as the last of April with great blessedness; without thunder or threatening, miles and miles of slant grey curtains drop between him and the outer world. Whether to lie out in it unfended and fireless is more or less distressful, is a matter of the point of view. A sheepman's fortune may depend on the number of days between lambing and shearing when the dormidera is too wet to unfold. It is a comfort in the heart of a hundred-mile spread of storm to sit under a canvas and notch these days as an augury on your staff.

Normally the parting of the flocks begins

immediately after shearing, but if possible the herders keep on in the valley until the lambs are weaned. This may occur at the end of about a hundred days and is best accomplished by a system of cross weaning, the lambs of one flock turned to the ewes of the next. But by whatever means, it is important to have older sheep with the young, so they become flock-wise and accustomed to the dogs. Not until all this has taken place are the flocks properly ready for the Long Trail, but before that the poppy gold which begins on the coastward fringes of the valley will have been cast well up on the slope of the Sierras, and about the centres of shepherd life begins to drift the first indubitable sign of a shearing, the smell of the Mexican cigarette.

III

 A SHEARING — THE CREW, THE CAMP, THE SHEARING BAILE, AND THE PARTING OF THE FLOCKS.

CHAPTER III

A SHEARING

To find a shearing, turn out from the towns
of the southern San Joaquin at the time of
the year when the hilltops begin to fray out
in the multitudinous keen spears of the wild
hyacinth, and look in the crumbling flakes of
the foothill road for the tracks of the wool
wagon. Here the roll of the valley up from
the place of its lagoons is by long mesas breaking into summits and shoulders; successive
crests of them reared up by slow, ample heav-

ings, settling into folds, with long, valleyward slopes, and blunt mountain-facing heads, flung up at last in the sharp tumult of the Sierras. Thereward the trail of the wool wagon bears evenly and white. Over it, preceded by the smell of cigarettes, go the shearing crews of swarthy men with good manners and the air of opera pirates.

When Solomon Jewett held the ranch above the ford by the river which was Rio Bravo, and is now Kern, shearings went forward in a manner suited to the large leisure of the time. That was in the early sixties, when there were no laborers but Indians. These drove the flocks out in the shoulder-high grasses; "for in those days," said Jewett, " we never thought feed any good, less than eighteen inches high," and at the week end rounded them up at headquarters for the small allowance of whiskey that alone held them to the six days' job. It was a condition of the weekly dole that all knives and weapons should be first surrendered, but as you can imagine, whiskey being hard to come by at that time, much water went to each man's flask; the nearer the bottom of the cask the more water.

" *No werito*, Don Solomon, *no werito*," complained the herders as they saw the liquor paling in the flasks, but it was still worth such service as they rendered.

The ration at Rio Bravo was chiefly atole or " tole " of flour and water, coffee made thick with sugar, and raw mutton which every man cut off and toasted for himself ; and a shearing then was a very jewel of the comfortable issue of labor. Of the day's allotment each man chose to shear what pleased him, and withdrawing, slept in the shade and the dust of the chaparral while his women struggled, with laughter and no bitterness of spirit, with the stubborn and over-wrinkled sheep. But even Indians, it seems, are amenable to the time, and I have it on the authority of Little Pete and the Manxman that Indians to-day make the best shearers, being crafty hand-workers and possessed of the communal instinct, liking to work and to loaf in company. Under the social stimulus they turn out an astonishing number of well-clipped muttons. Round the half moon of the lower San Joaquin the Mexicans are almost the only shearers to be had, and even the men

who employ them credit them with the greatest
fertility in excuses for quitting work.

All the lost weathers of romance collect
between the ranges of the San Joaquin, like
old galleons adrift in purple, open spaces of
Sargasso. Shearing weather is a derelict from
the time of Admetus; gladness comes out of
the earth and exhales light. It has its note,
too, in pipings of the Dauphinoises, seated on
the ground with gilias coming up between
their knees while the flutes remember France.
Under the low, false firmament of cloud, pools
of luminosity collect in interlacing shallows of
the hills. Here in one of those gentle swales
where sheep were always meant to be, a ewe
covers her belated lamb, or has stolen out from
the wardship of the dogs to linger until the
decaying clot of bones and hide, which was
once her young, dissolves into its essences. The
flock from which she strayed feeds toward the
flutter of a white rag on the hilltop that signals a shearing going on in the clear space of
a cañon below. Plain on the skyline with his
sharp-eared dogs the herder leans upon his
staff.

As many owners will combine for a shearing as can feed their flocks in the contiguous pastures. At Noriega's this year there were twenty-eight thousand head. Noriega's camp and corrals lie in the cañon of Poso Creek where there is a well of one burro power, for at this season the rains have not unlocked the sources of the stream. Hills march around it, shrubless, treeless; scarps of the Sierras stand up behind. Tents there are for stores, but all the operations of the camp are carried on out of doors. Confessedly or not, the several sorts of men who have to do with sheep mutually despise one another. Therefore the shearing crew has its own outfit, distinct from the camp of the hired herders.

Expect the best cooking and the worst smells at the camp of the French shepherds. It smells of mutton and old cheese, of onions and claret and garlic and tobacco, sustained and pervaded by the smell of sheep. This is the acceptable holiday smell, for when the far-called flocks come in to the shearing then is the only playtime the herder knows. Then if ever he gets a blink at a pretty girl, claret,

and *bocie* at Vivian's, or a game of hand-ball
at Noriega's, played with the great shovel-

shaped gloves that
are stamped with
the name of Pam-
plona to remind
him of home. But
by the smell chiefly
you should know
something of the
man whose camp you have come on unawares.
When you can detect cheese at a dozen yards
presume a Frenchman, but a leather wine bottle proves him a Basque, garlic and onions
without cheese, a Mexican, and the absence of
all these one of the variable types that calls
itself American.

The shearing sheds face one side of the
corrals and runways by which the sheep are
passed through a chute to the shearers. The
sheds, of which there may be a dozen, accommodate five or six shearers, and are, according
to the notion of the owner, roofed and hung
with canvas or lightly built of brush and
blanket rags. Outside runs a shelf where the

packers tie the wool. One of them stands at every shed with his tie-box and a hank of tie-cord wound about his body. This tie-box is merely a wooden frame of the capacity of one fleece, notched to hold the cord, which, once adjusted, can be tightened with a jerk and a hitch or two, making the fleece into a neat, square bundle weighing six to ten pounds as the clip runs light or heavy. Besides these, there must go to a full shearing crew two men to handle the wool sacks and one to sit on the packed fleeces and keep tally as the shearer cries his own number and the number of his sheep, betraying his country by his tongue.

"*Numero neuf, onze!*" sings the shearer.

"*Numero neuf, onze!*" drones the marker.

"*Cinco; viente!*"

"*Numero cinco; viente!* tally."

I have heard Little Pete keep tally in three languages at once.

The day's work begins stiffly, little laughter, and the leisurely whet of shears. The pulse of work rises with the warmth, the crisp bite of the blades, the rustle and scamper of sheep in the corral beat into rhythm with the bent backs rising and stooping to the incessant cry, " *Numero diez, triente !* " " Number ten, tally!" closing full at noon with the clink of canteens. Afternoon sees the sweat dripping and a freer accompaniment of talk, drowned again in the rising fever of work at the turn of the day, after which the smell of cooking begins to climb above the smells of the cor

rals. A man wipes his shears on his overalls and hangs them up when he has clipped the forty or fifty sheep that his wage, necessity, or his reputation demands of him.

Two men can sack the wool of a thousand sheep in a day, though their contrivances are the simplest,—a frame tall enough to be taller than a wool sack, which is once and a half as tall as Little Pete, an iron ring over which the wetted mouth of the

sack is turned and so held fast to the top of
the frame, a pole to support the weight of the
sack while the packer sews it up. Once the
sack is adjusted, with ears tied in the bottom
corners over a handful of wool, the bundled
fleeces are tossed up into it and trampled close
by the packer as the sack fills and fills. The
pole works under the frame like an ancient
wellsweep, hoisting the three hundred pound
weight of wool while the packer closes the
top.

For the reason why wool shears are ground
dull at the point, and for knowing about the
yolk of the wool, I commend you to Noriega
or Little Pete; this much of a shearing is their
business; the rest of it is romance and my
province.

The far-called flocks come in; Raymúndo
has climbed to the top of the wool sack tower
and spies for the dust of their coming; dust
in the east against the roan-colored hills; dust
in the misty, blue ring of the west; high dust
under Breckenridge floating across the banked
poppy fires; flocks moving on the cactus-grown
mesa. Now they wheel, and the sun shows them

white and newly shorn; there passes the band
of Jean Moynier, shorn yesterday. Northward
the sagebrush melts and stirs in a stream of
moving shadow.

"That," says Raymúndo, "should be
Étienne Picquard; when he goes, he goes fast;
when he rests, he rests altogether. Now he
shall pay me for that crook he had of me last
year."

"Look over against the spotted hill, there
by the white scar," says a little red man who
has just come in. "See you anything?"

"Buzzards flying over," says Raymúndo
from the sacking frame.

"By noon, then, you should see a flock
coming; it should be White Mountain Joe. I
passed him Tuesday. He has a cougar's skin,
the largest ever. Four nights it came, and
on the fourth it stayed."

So announced and forerun by word of their
adventures the herders of the Long Trail
come in. At night, like kinsmen met in hos-
telries, they talk between spread pallets by the
dying fires.

"You, Octavieu, you think you are the only

one who has the ill fortune, you and your poisoned meadows! When I came by Oak Creek I lost twoscore of my lambs to the forest ranger. Twoscore fat and well grown. We fed along the line of the Reserve, and the flock scattered. Ah, how should I know, there being no monuments at that place! They went but a flock length over, that I swear to you, and the ranger came riding on us from the oaks and charged the sheep; he was a new man and a fool not to know that a broken flock travels up. The more he ran after them the farther they went in the Reserve. Twoscore lambs were lost in the steep rocks, or died from the running, and of the ewes that lost their lambs seven broke back in the night, and I could not go in to the Reserve to hunt them. And how is that for ill fortune? You with your halfscore of scabby wethers!"

Trouble with forest rangers is a fruitful topic, and brings a stream of invective that falls away as does all talk out of doors to a note of humorous large content. Jules upbraids his collie tenderly:—

"So you would run away to the town, eh,

and get a beating for your pains; you are well served, you misbegotten son of a thief! Know you not there is none but old Jules can abide the sight of you?"

Echenique by the fire is beginning a bear story:—

" It was four of the sun when he came upon me where I camped by the Red Hill northward from Agua Hediónda and would have taken my best wether, Duroc, that I have raised by my own hand. I, being a fool, had left my gun at Tres Pinos on account of the rangers. Eh, I would not have cared for a sheep more or less, but Duroc!—when I think of that I go at him with my staff, for I am seven times a fool, and the bear he leaves the sheep to come after me. Well I know the ways of bears, that they can run faster than a man up a hill or down; but around and around, that is where the great weight of Monsieur le Bear has him at fault. So long as you run with the side of the hill the bear comes out below you. Now this Red Hill where I am camped is small, that a man might run around it in half an hour. So I run and the bear runs; when I come out again by my sheep I speak to the dogs that they keep them close. Then I run around and around, and this second time — Sacre ! "

He gets upon his feet as there rises a sudden scurry from the flock, turned out that evening from the shearing pens and bedded on the mesa's edge, yearning toward the fresh feed. Echenique lifts up his staff and whistles to his dogs; like enough the flock will move out in the night to feed and the herder with him. Not until they meet again by chance, in the summer meadows, will each and several hear the end of the bear story. So they recount the year's work by the shearing fires, and if they be hirelings of different owners, lie to each other about the feed. Dogs snuggle to their masters; for my part I believe they would take part in the conversation if they could, and suffer in the deprivation.

At shearings flocks are reorganized for the Long Trail. Wethers and non-productive ewes are cut out for market, yearlings change hands, lambs are marked, herders outfitted. The shearing crew which has begun in the extreme southern end of the valley passes north on the trail of vanishing snows even as far as Montana, and picks up the fall shearings, rounding toward home. This is a recent procedure. Once there was time enough for a *fiesta* lasting two or three days, or at the least a shearing *baile*. I remember very well when at Adobe, before the wind had cleared the litter of fleeces, they would be riding at the ring and clinking the shearing wage over cockfights and monte. Toward nightfall from somewhere in the blue-and-white desertness, music of guitars floated in the prettiest girls in the company of limber vaqueros, clinking their spurs and shaking from their hair the shining crease where the heavy sombrero had rested. Middle-aged señoras wound their fat arms in their rebosas and sat against the wall; blue smoke of cigarettes began to sway with the strum of the plucked guitar; cascarones would fly about, breaking in bright tinsel showers. O, the sound of the mandolin, and the rose in the señorita's hair! *What* is it in the Castilian strain that makes it possible for a girl to stick a rose behind her ear and cause you to forget the smell of garlic and the reek of unwashed walls?

Along about the middle hours, heaves up, heralded by soft clinkings and girding of broad tires, the freighter's twenty-eight-mule team. The teamsters, who have pushed their fagged animals miles beyond their daily stunt to this end, drop the reins to the swamper and whirl with undaunted freshness to the dance. As late as seven o'clock in the morning you could still see their ruddy or freckled faces glowing above the soft, dark heads. Though if you had sheep in charge you could hardly have stayed so long. Outside so far that the light that rays from the crevices of the bursting doors of Adobe is no brighter than his dying fire, the herder lies with his sheep, and by the time the bleached hollows of the sands collect shadows tenuous and blue, has begun to move his flock toward the much desired Sierra pastures.

IV

THE HIRELING SHEPHERD —
WITH SOME ACCOUNT OF HOW HE HAS BECOME AN ABOMINATION, AND OF THE MEN WHO HIRE HIM

CHAPTER IV

THE HIRELING SHEPHERD

"AND now," says the interlocutor, "tell me what led you first to this business of sheep?"

That was at Little Pete's shearing at Big Pine, a mile below the town ; a wide open day of May, dahlia coming into bloom and blue gilias quavering in the tight shadows under the sage. Pete had been showing me the use of a shepherd's crook, not nearly so interesting as it sounds. He hooked it under the hind leg of a wether and drew him into the shearing

pen; now he leaned upon its long handle as
on a staff.

"In Arles where I was born, by the Rhone,"
said Pete, "my father kept sheep."

"And you were put to the minding of
them?"

"As a boy. We drove them to the Alps in
summer, I remember very well. We went between the fenced pastures, feeding every other
day and driving at night. In the dark we
heard the bells ahead and slept upon our feet.
Myself and another herd boy, we tied ourselves together not to wander from the road.
We slept upon our feet but kept moving to the
bells. This is truth that I tell you. Whenever
shepherds from the Rhone are met about
camps in the Sierras they will be talking of
how they slept upon their feet and followed
after the bells."

There was a clump of crimson mallow at
the corner of the shearing corral. I remembered what the Indians had told me in this
sandy waste, that where the mallow grew they
digged and found, if no more, at least a handful of plastic clay for making pots. That was

like any statement of Pete's; if you looked for
it, there was always a good lump of romance
about its roots.

" All that country about the Rhone," he
said, " is of fields and pastures, and the Alps
hang above them like clouds. Meadows of the
Sierras are green, but not so green as the little
fields of France when we went between them
with the flocks. We fed for three months in
the high pastures, and for idleness wove gar-
ters in curious patterns of woolen thread, red
and green and blue. Yes; for our sweethearts,
they wore them on holidays. But here it seems
a garter is not to be mentioned."

" And you came to America ? "

" Yes; there were changes, and I had heard
that there was free pasture, and money — Eh,
yes, it passes freely about, but there is not
much that sticks to the fingers." Pete shunted
the dodge-gate in the pens and searched the
horizon for the dust of his flocks.

" And you, Enscaldunac ? "

The Basco lifted his shoulders and folded
his arms above his staff.

" In the Pyrenees my father keep sheep,

his father keep sheep, his father"— He threw
out his hands inimitably across the shifting
shoulders of the flock; it was as if he had directed the imagination over a backward stretch
of time, that showed to its far diminishing end
generations of small, hairy men, keeping
sheep.

"It is soon told," said Sanger, his voice
halting over some forgotten burr of speech,
"how I began to be interested in sheep.

"It was in Germany when I was a boy.
Every man has two or three head in his stable,
and there will be one herd boy to the village;
he leads them out to feed, and home at night.
Every sheep knows its own fold. They are
like dogs returning to the doorstep when they
come in at night, and in the morning they
bleat at the voice of the herd boy. But here
we run two and three thousand to the flock."

The Manxman, when the question was put
to him, laid the tips of his thin fingers together
deliberatively, between his knees.

"Well, I began working a shearing crew,
my brother and I, but, you see, in the Isle o'
Man"— What more would you have? Once

a man has been put to the care of sheep he reverts to it in any turn of his affairs like mavericks to old water holes. And if he would keep out of the business, he must keep strictly away from the smell of the dust they beat up on the trail and the familiar blether of the flock. Narcisse Duplin, who used regularly to damn the business in October and sell out, and as regularly buy again in February, told me this, and told at the same time of a certain banker in an inland town who had made his money in sheep and was now ashamed of it, who kept a cosset ewe in his back yard. There used to be at Tres Pinos a man who had sold two thousand wethers and a thousand ewes, to buy a little shop where he could sell lentils and claret and copper-riveted overalls to the herders going by on the Long Trail. But he never came to any good in it, for the reason that when trade should be busiest at the semi-annual passage of the flocks, he would be out walking after the sheep in the smell and the bitter dust.

That most sheep-herders are foreigners accounts largely for the abomination in which

they are held and the prejudice that attaches
to the term. American owners prefer to be
called wool growers, but it is well to be exactly
informed. The Frenchmen call themselves
bergers, the Mexicans *boregeros*, the Basques
artzainas, of all which shepherd is the exact
equivalent. Sheep-herder is a pure colloquial-
ism of the man outside and should not be made
to stand for more than it includes. The best
terms of a trade are to be found among the
men who live by it, and these are their proper
distinctions: The owner or wool grower sits at
home, and seldom seeing his flocks sends them
out under a head shepherd or major-domo; a
shepherd is an owner who travels with the
flock, with or without herders, overseeing and
directing; the sheep-herder is merely a hireling who works the flock in its year-long passage from shearing to shearing.

This is the first estate of most sheepmen.
The herder runs a flock for a year or two for
a daily wage of tobacco and food and a dollar, and if he has no family, fifty dollars is as
much as he finds occasion to spend upon himself. Then he takes pay in a bunch of ewes

and runs them with his master's flock. With
the year's increase he unites with some other
small owner, and puts his knowledge of pastures to the proof. After this his affairs are in
the hands of the Little Gods of Rain. Three or
four successive dry years return him " broke "
to the estate of herding ; the same number of
years of abundant wetness make him a wool
grower.

Notable owners, such as Watterson, Olcese,
Sanger, and Harry Quinn of Rag Gulch, think
themselves not much occupied with romance.
They improve the breeds, conserve the natural
range, multiply contrivances. At Rag Gulch
there is a cemented vat for dipping sheep, and
at Button Willow they have set up wool-clip-
ping machines, — but as for me, the dust of
the shuffling hoofs is in my eyes. As it rises
on the trail one perceives through its pale
luminosity the social order struggling into
shape.

Sanger, when he drove his sheep to Montana in '70, went up like a patriarch with his
family in wagons, his dogs and his herders,
his milch cows, his saddle horses, and his sheep

in bands. When they came by living springs,
there they pitched the camp; when they found
fresh pastures, there they halted. But on the
Long Trail the herders go out with a little
burro to pack, with a lump of salt pork and

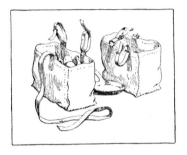

a bag of lentils, a
bunch of garlic,
a frying pan, and
a pot, with two or
three dogs and a
cat to ride on top of
the cayaques and
clear the camp of
mice. After them comes the head shepherd in
a stout-built wagon. Met on the county roads,
he is to be distinguished from the farmers by
the sharp noses of the dogs thrust out between
his feet, and by the appearance of having on
too many clothes and the clothes not belong-
ing to him. Nothing sets so ill on the man
from outdoors as the ready-made suit. On the
range in a blouse loose at the throat, belted
with a wisp of sheepskin or a bright handkerchief, these shepherd folk show to be admi-
rably built, the bodies columnar, the chests

brawny, the reach of the arms extraordinary, the hands not calloused but broadened at the knuckles by the constant grip of the staff.

Of the other sorts of men having to do with sheep there are not many who merit much attention. These are the buyers who seek out the flocks on the range, and fortified by a secret knowledge of the market fluctuations, bargain for the mutton and the fleeces. Having paid to the shepherd, as earnest of their intention, the cost of driving the flock at a given time to the point of transportation, they melt away by the main traveled roads, and the herder knows them no more. The real focus of the sheep business in any district is to be found in some such friendly concern as the house of Olcese and Ardizzi, who make good in the terms of modernity the very old rule that one Frenchman is always worth being trusted by another. Hardly any who go up across my country but have been lifted by them through their bad years by credits and supplies, and the inestimable advantage that comes to a man in knowing his word is esteemed good.

Once for all the French herders in America

shall have in me a faithful recorder. You may call a Frenchman a Gascon, which is to say a liar, and escape punishment; but you really must not confound him with a Basque. Understand that all the Pyreneeans of my acquaintance are straight folk and likable, but if you lay all the evils of shepherding at the doors of those I do not know, you will have some notion of how they are esteemed of the French.

When on the mesa or about the edges of a gentian-spattered meadow you come upon a still camp with "Consuelo," the "Fables of La Fontaine," or Michelet's "Histoire de France" lying about among the cooking pots, it is well to wait until the herder comes home. In seventeen years I have found nobody better worth than Little Pete to discuss French literature. This is that Pierre Geraud who has the meadow of Coyote Valley and the ranch at Tinnemaha; a man who gives the impression that he has made himself a little less than large for convenience in getting about, of such abundant vitality and elasticity that he gives back largely to the lightest touch. He knows

how to put information in its most pregnant
shape, though I am not sure it is because he
is a Frenchman or because he is a shepherd.

Once you get speech with them, of all outdoor folk the minders of flocks are the most
fruitful talkers; better at it than cowboys, next
best after forest rangers. The constant flux
from the estate of owner to hireling makes
them philosophers; all outdoors contrives to
nourish the imagination, and they have in full
what we oftenest barely brush wings with, elemental human experiences.

Once in the Temblors, a wild bulk of hills
westward from San Emigdio, I knew a herder
who had called a woman from one of the wattled huts sprawled in a brown cañon; she an-
swering freely to the call as the quail to the
piping of its mate. She was slim and brown,
and points of amber flame swam in her quiet
eyes. They went up unweariedly by faint old
trails and felt the earth-pulse under them.
They shook the unregarded rain from their
eyes, and sat together in a wordless sweet companionship through endless idle noons. Afterward when she grew heavy he set her Madonna-

wise on a burro, he holding the leading strap and she smiling at him in a large content.

Well — but what *is* marriage exactly?

Understand that the actual management of a flock on the range is never a " white man's job." Those so describing themselves who  may be hired to it are the impossibles, men who work a little in order to drink a great deal, returning to the flock in such a condition of disrepair that their own dogs do not know them.

Of the twoscore shepherds who pass and repass between Naboth's field and the foot of Kearsarge, most are French, then Basque, Mexican, and a Portuguese or two. Once I found a Scotchman sitting on a fallen plinth

of the Black Rock below Little Lake; I knew
he was Scotch because he was knitting and
he would not talk. There was an Indian who
worked for Joe Espelier, — but in general the
Indian loves society too much to make a notable herder, and the Mexican has a difficulty in
remembering that the claims of his employer
are superior to the obligations of hospitality.
Gervaise told me that when he ran thirty thousand merinos in New Mexico he used to deal
out supplies in day's rations, otherwise he
would be feeding all his herders' relations and
relations-in-law.

It is said of the Devil that he spent seven
years in learning the Basque language and
acquired but three words of it, and offered in
corroboration that the people of the Pyrenees
called themselves Enscaldunac, "the people
with a speech." I believe myself these Bascos
are a little proud of the foolish gaspings and
gutterings by which they prevent an understanding, and contribute to the unfounded assumption that most sheep-herders are a little
insane. This sort of opprobrium is always cast
upon unfamiliar manners by the sorts of peo-

ple who meet oftenest with shepherd folk, —
cowboys, homesteaders, provincials with little
imagination and no social experience. Whenever it is possible to bridge the prejudice which
isolates the herder from the servants of other
affairs, what first appears is that the grazing
ground is the prize of a little war that requires
for its successful issue as much foresightedness
and knowledge of technique as goes propor-
tionately to other business, so that a man
might much more easily go insane under its
perplexities than for the want of employment
that is oftenest imputed. Nor does shepherding lack a sustaining morale in the occasions
it affords for devotion to the interests of the
employer. And this presents itself in any
knowledgeable report of their relations that,
in a business carried on so far from the owner's eye, nothing could be possible without an
extraordinary degree of dependableness in the
hireling.

Not that the leash of reason does not occasionally slip in the big wilderness; there was
Jean Lambert, who in a succession of dry years
found himself so harassed by settlers and cattle-

men occupying his accustomed ground and defending them with guns and strategies, that he conceived the very earth and sky in league against him, and was found at last roaring about a dry meadow, holding close his starved flock and defying the Powers of the Air. Once there was a Portuguese herder misled by false monuments in the Coso country, without water for three days, discovered witless and happy, bathing nakedly in the waters of mirage. But there were also miners in that county and teamsters whom the land made mad; indeed, what occupation fends us from thirst and desertness? I hand you up these things as they were told to me, for such as these always occur in some other place, like Arizona or New Mexico where almost anything might happen. With all my seeking into desert places there are three things that of my own knowledge I have not seen, — a man who has rediscovered a lost mine, the heirs of one who died of the bite of a sidewinder, and a shepherd who is insane.

The loneliness imputed by the town-bred is not so in fact. Almost invariably two men are put to a flock, and these are seldom three days together out of touch with the owner or head shepherd who, traveling with supplies, directs several bands at once, baking bread, replenishing the outfit, spying ahead for fresh pastures, and purveying news. This necessity for renewing contact at given places and occasions points the labor of the herder and supplies a companionable touch. Herders of different owners meet on the range and exchange misinformation about the feed; lately also they defame the forest rangers. Returning in the fall, before undertaking the desert drive, they turn into the alfalfa fields about Oak Creek and below Williamson and Lone Pine. Here while the flock fattens they make camps of ten or a dozen; here in long twilights they sing and romp boyishly with the dogs, and here the wineskin goes about. These goatskin bottles with the hair inside come from Basqueland and are held by the possessors to give an unrivaled flavor to the weak claret drunk in camp. When a company of Basque herders are met about the fire, in the whole of a long evening the wineskin does not touch the ground. Each man receives it from his neighbor, holds it a foot away from

his face, deftly wets his throat with a thin, pink
stream squirted through the horn tip, hands it
about and about, singing.

After sundown in the stillness of high valleys
the sound of an accordion carries far. When
it croons wheezily over a love song of the sev-
enteenth century, it is worth following to its
point of issue beside the low flare of the brushwood fire with the shepherds seated round it
on the ground. There you will hear roundels
and old ballades, perhaps a new one beginning, —

> "A shepherd there was of Gascony,
> A glutton, a drunkard, a liar was he,
> A rascal, a thief, and a Blasphemer,
> The worst in the whole round world I aver ;
> Who, seeing the master had left him alone,
> He gave the coyotes the lambs for their own,
> He left the poor dogs to watch over the sheep
> And down by the wine cask he laid him asleep."

It goes much more swingingly than that in
the original, which, if you wish, you can get
from Little Pete, who made it.

V

THE LONG TRAIL —
HOW IT WAS DEFINED,
WHAT GOES ON IN IT,
AND HOW THE DAY'S
WORK IS ACCOMPLISHED.

CHAPTER V

THE LONG TRAIL

TOWARD the end of spring in the wide California valleys, night begins close along the ground, as if it laired by day in the shadows of the rabbit-brush or suspired sleepily from thick, secret sloughs. At that hour when the earth turns as if from the red eye of the sun, all the effort of nature seems to withdraw attention from its adumbration to direct it toward the ineffably pure vault of blueness on which the clear obscurity that shores the rim of the world encroaches late or not at all. In the San Joaquin there will be nights of early summer when the live color of heaven is to be seen at all hours beyond the earth's penumbra, darkling between the orderly perspectives of the stars. At such seasons there will be winking in the pellucid gloom, in the vicinity of shearing stations, a hundred camp fires of men who have not lost the sense of the earth being good to lie down upon. They have moved out from Famoso, from Delano, Poso, and Caliente, bound as the mind of the head shepherd runs for summer pastures as far north as may be conveniently accomplished between shearing and lambing; and all the ways of their going and coming make that most notable of sheepwalks, the Long Trail.

The great trunk of the trail lies along the east slope of the Sierra Nevadas, looping through them by way of the passes around Yosemite, or even as far north as Tahoe, shaped and defined by the occasions that in little record the progress from nomadism to the commonwealth. Conceive the cimeter blade of the Sierra curving to the slow oval of the valley, dividing the rains, clouds herding about its summits and flocks along its flanks, their approaches ordered by the extension and recession of its snows. The common necessities of the sheep business beat it into a kind of rhythm as early even as the time when every foot of this country was open range. Recurrently as the hills clothed themselves with white wonder the shepherds turned south for lambing, and as surely as bent heather recovers from the drifts, they sought the summer pastures.

The down plunge of the Sierras to the San Joaquin is prolonged by round-backed droves of hills, and the westerly trail is as wide as a week of flock journeys; but here on the east you have the long, sharp scar where *Padahoon*, the little hawk who made it, tore the range from its foundations when he stole that territory from the little duck who brought up the stuff for its building from the bottom of the primordial sea. Here the trail hugs the foot of the great Sierra fault for a hundred miles through the knife-cut valleys, trending no farther desertward than the scant fling of winter rains, and even here it began soon enough to be man-crowded.

Wherever the waters of cloud-dividing ridges issue from the cañons, steadying their swaying to the level lands, there were homesteads established that in thirty years expanded into the irrigated belt that limits and defines the range of sheep. Not without a struggle though. Between the herders and the ranchers the impalpable fence of the law had first to externalize itself in miles upon miles of barbed wire to accomplish for the patented lands what the hair rope is supposed to do for the teamster's bed, for in the early eighties there was no vermin so pestiferous to the isolated rancher as the sheep. Finally the trail was mapped by the viewless line of the Forest Reserve, drawn about the best of the watershed and so narrowed that where it passes between Kearsarge and Naboth's field, where my house is, it is no more than a three-mile strip of close-grazed, social shrubs.

The trail begins properly at the Place of the Year Long Wind, otherwise Mojave. Flocks pour into it by way of Tehachapi, and in very dry years from as far south as San Gabriel and San Bernardino, crowded up with limping, stark-ribbed cattle. In the spring of '94 they were

driven north in such numbers that the stage
road between Mojave and Red Rock was trod-
den indistinguishably into the dust. The place
where it had been was mapped in the upper air
by the wide, tilted wings of scavengers and the
crawling dustheaps below them on the sand,
formless blurs for the sheep and long snaking
lines of steers ; for horned cattle have come
so much nearer the man-mind that they love
a beaten path. Weeks
on end the black gui-
dons flapped and halt-
ed in the high currents
of the furnace-heated
air.

Rolling northward
on the Mojave stage,
from the high seat be-
side the driver, I saw
the sick hearts of cat-
tlemen and herders
watch through swollen
eyelids the third and then the half of their
possessions wasting from them as sand slips
through the fingers. By the dry wash where

they buried the Chinaman who tried to walk in from Borax Marsh without water, we saw Baptiste the Portuguese, sitting with his eyes upon the ground, all his flock cast up along the bank, and his hopes with them like the waste of rotting leaves among the bleached boulders of a vanished stream, dying upon their feet.

All trails run together through Red Rock, the gorge by which the stage road climbs to the mesa. There is a water hole halfway of its wind-sculptured walls; often had I seen it glimmering palely like a dead eye between lashless, ruined lids. Crowded into the defile at noon, for at that time we made the first stage of the journey by day, a band of black faces added the rank smell of their fleeces to the choked atmosphere. The light above the smitten sands shuddered everywhere with heat. The sheep had come from Antelope Valley with insufficient feed and no water since Mojave, and had waited four hours in the breathless gully for the watering of a band of cattle at the flat, turgid well. The stage pushed into the cañon as having the right of way, for besides passengers we carried the mail; the herder spoke to the dogs that they open the flock to let us pass. They and the sheep answered heavily, being greatly spent; dumbly they shuffled from the road and closed huddling behind, as clods. For an interval we halted in the middle of the band until one of the horses snorted back upon his haunches and occasioned one of those incidents that, whether among sheep or men, turn us sickeningly from the social use of the flock-mind. The band began to turn upon itself; those scrambling from the horses piled up upon their fellows as viewless shapes of thirst and fear herded them inward to the suffocating heap that sunk and shuddered and piled again. My eyes were shut, but I heard the driver swear whispering and helplessly for the brief interval that we could not hear the gride of the moving wheels upon the sand. Afterward when I came to my own place I watched the trail long for the passing of that herder and that band, to inquire how they had come through, — *but they never passed!*

Nothing, absolutely nothing, say the herders, of interest or profit can happen to a flock be-

tween Antelope Valley and Haiwai in a dry year.
It is the breeding place of little dust devils that
choose the moment when your pot lid is off, or
you cool your broth with your breath, to whisk
up surprisingly out of stillness with rubbish
and bitter dust to disorder the camp. Footsoreness, loco-weed, deadly waters, and starvation establish its borders ; and withal no possibility of imputing malignity. It is not that the
desert would destroy men and flocks, it merely
neglects them. When they fail through its
sheer inattention, because of the preoccupation
of its own beauty, it has not time even to kill
quickly. Plainly the lord of its luminous great
spaces has a more tremendous notion, not to
be disturbed for starveling ewe, not though
the bloomy violet glow of its twilight closes
so many times on the vulture dropped above
it, swinging as from some invisible pendulum
under the sky. Lungren showed me a picture
once, of a man and a horse dead upon the
desert, painted as it would be with the light
breaking upon the distended bodies, nebu-
lously rainbow-hued and tender, which he said
hardly anybody liked. How should they ? It is

still hard for men to get along with God for thinking of death not as they do.

But if ever spring comes to the Mojave, and the passage of spring beyond the Sierra wall is a matter of place and occasion rather than season, there is no more tolerable land for a flock to be abroad in. This year it came and stayed along three hundred miles, and the sheep grew fat and improved their fleeces. But for the insufficience of watering places a hundred thousand might have thriven on the great variety of grazing, — atriplexes, dahlia, tender young lupines, and " marrow-fat " weed.

As many shepherds as think the grudging permission to cross the Forest Reserve not too dearly paid for by the vexations of it, bring their sheep up by way of Havilah and Green-horn through Walker's Pass. As many as think it worth while feed out toward Panamint and Coso, where once in seven years there is a chance of abundant grazing; but about Owen's Lake they are drawn together by the narrowing of the trail and the tax collector. If ever you come along the south shore of that dwindling, tideless water about the place where Manuel

de Borba killed Mariana, his master, and sold
the flock to his own profit, look across it to the
wall-sided hulks of the Sierras ; best if you can
see them in the pure, shadowless light of early
evening when the lake shines in the wet grey
color of Irish eyes. For then and from this
point it seems the Indians named them " *Too-
rápe*," the Ball Players. They line up as braves
for the ancient play, immortally young, shining
nakedly above, girt with pines, their strong
cliffs leaning to the noble poises of the game.

" It is evident," Narcisse Duplin used to say
when he came to this point, " that God and a
poor shepherd may admire the same things."

Always in October or April one sees about
the little towns of Inyo, in some corner of the
fields, two to six heavy wagons of the head shepherds, with the season's outfit stowed under
canvas ; and at Eibeshutz's or Meysan's happen upon nearly unintelligible herders buying
the best imported olive oil and the heaviest
American cowhide boots. Hereabouts they
refresh the trail-weary flocks in the hired pastures and outfit them for the Sierra meadows.
Here also they pay the license for the open

range, two to five cents a head, payable by
actual count in every county going or returning. As the annual passage is often twice
across three or four counties, the license becomes, in the minds of some herders, a thing
worth avoiding. Narcisse Duplin, red Narcisse,
who went over this trail once too often, told
me how, in a certain county where the land
permitted it, he would hide away the half of
his flock in the hills, then go boldly with the
remnant to pay his assessment, smuggling forth
the others at night out of the collector's range.
But here where the trail spindles out past
Kearsarge there is no convenience and, I may
add, hardly any intention of avoiding it.

A flock on the trail moves out by earliest
light to feed. For an hour it may be safely
left to the dogs while the herder starts the fire
under his coffee pot and prepares his bowl of
goat's milk and large lumps of bread. The
flock spreads fanwise, feeding from the sun.
Good herding must not be close; where the
sheep are held in too narrow a compass the
middlers and tailers crop only stubble, and

coming empty to the bedding ground, break
in the night and stray in search of pasture.
An anxious herder makes a lean flock. Properly the band comes to rest about mid-morning,
drinking when there is water to be had, but if
no water, ruminating contentedly on the open
fronts of hills while the herder cooks a meal.

Myself, I like the dinner that comes out of
the herder's black pot, mixing its savory smells
with the acrid smoke of burning sage. You
sit on the ground under a little pent of brush
and are served in a tin basin with mutton, lentils, and garlic cooked together with potatoes
and peppers ("red pottage of lentils"), with
thick wedges of sour-dough bread to sop up
the gravy, good coffee in a tin cup; and after
the plate is cleared, a helping of wild honey
or tinned sweet stuff. Occasionally there will
be wild salad, miner's lettuce, pepper grass or
cress from springy meadows. If the herder has
been much about Indians, you may have little
green pods of milkweed cooked like string
beans, summers in westward-fronting cañons,
thimbleberries which the herder gathers in his
hat. Trout there are in a trout country, but

seldom game, for a gun does not go easily in a cayaca.

When in the fall the Basques forgather at a place on Oak Creek called by the Indians "*Sagaharawite*, Place-of-the-Mush-that-was-Afraid," you get the greatest delicacy of a sheep camp, a haunch of mutton stuck full of garlic corns and roasted in a Dutch oven under ground. Even buried a foot in red-hot coals the smell of this delectation is so persuasive that Julien told me once on Kern River, when he had left his mutton a moment to look after the sheep, a bear came out of the hills and carried off the roast in the pot. There is no doubt whatever of the truth of this incident.

Bread for the camp is baked by the head shepherd, and when it is ready for the pans he

pulls off a lump and drops it back in the flour sack. There it ferments until it is used to start the next baking.

"How long," said I to the herder from whom I first learned the management of the loaves, "how long might you go on raising bread from one 'starter'?"

He considered as he rubbed the dough from his hands.

"When first I come to this country in '96 I have a fresh piece, from the head shepherd of Louis Olcese. Yes, when I am come from France. Madame-who-writes-the-book could not have supposed that I brought it with me. Ah, *non !* "

A sack of flour goes to six of the round, brown loaves, and one is a four days' ration, excellent enough when it comes up out of the baking trench, rather falling off after three days in the pack with garlic and burro sweat, and old cheese. The acceptable vegetables are lentils and onions, and the test of a good employer is the quantity of onions that can be gotten out of him after the price goes higher than a dollar and a quarter a sack.

The mess which the herder puts over the fire every day at mid-morning is packed in the pot in the cayaca when the flock moves out in the afternoon, and warmed at his twilight-cheering fire, serves as supper for himself and the dogs alike, and not infrequently in the same dish.

I have said you should hear what the tariff revision accomplished for the sheep. Just this: before that, men raised sheep for wool or mutton expressly, but chiefly for wool. Then as the scale of prices hung wavering, doubtful if wool or mutton was to run highest, they began to cross the wool and mutton breeds to produce a sheep that matures rapidly and shears nine or ten pounds of wool, directing the management of the flock always towards the turn of the highest prices. Every sheepman will have his preferences among Merinos, Shropshires, and Cotswolds; but in general the Merinos are most tractable, and blackfaces the best for fenced pastures, for though they are marketable early they scatter too much, not liking to feed in the middle of the band, grow footsore too easily, and despise the herder. It is the ultimate disposition of the flocks, whether for mutton or wool, that determines the distribution of them along the upper country contiguous to the trail, as the various sorts of forage, in the estimation of the shepherd, favor one or another end. He is a poor shepherd whose mind cannot outrun the flock by a season's length when by eight and nine mile journeyings they pass northward in the spring. Little Pete drops out at Coyote Valley where by owning the best meadow he controls the neighboring feed. Joe Eyraud, White Mountain Joe, turns off toward the upswelling of his name peak to the perennial pastures of its snows. One goes by Deep Springs and Lida to the far-between grazing-grounds of Nevada, another to the burnt desert of Mono. Time was before the Forest Reserve cut them off from the high Sierras, the shepherds worked clean through them, returning to the lambing stations by way of North Fork, Kaweah, and the Four Creek country, and such as came up the west slope went back through Mono and Inyo. But now they return as they went, complaining greatly of depleted pastures.

The flocks, I say, drift northward where the turgid creeks discharge on the long mesas. Passage toward the high valleys is deterred by late melting of the snows and urged forward by the consideration that along the most traveled stages of the way there will be no new feed between the flowering of wild almonds and the time of Bigelovia bloom. Close spring feeding makes a bitter passage of the fall returning. In bad years the flocks turn in to the barley stubble, they take the last crop of alfalfa standing; in a vineyard country they are put to stripping the leaves from the vines.

What the shepherd prays for when in the fall the tall dust columns begin to rise from the Black Rock is a promise of rain in the dun clouds stretched across the valley, low and fleecy soft, touching the mountains on either side; grey air moving on the dusky mesas, wide fans of light cutting through the cañons to illume the clear blue above the Passes; soft thunder treading tiptoe above the floor of cloud, moving about this business of the rain.

VI

THE OPEN RANGE — THE COUNTRY WHERE THERE IS NO WEATHER, AND THE SIERRA MEADOWS

CHAPTER VI

THE OPEN RANGE

Beyond that portion of the great California sheepwalk which is every man's, the desert-fenced portion between Mojave and Sherwin Hill, lies a big, wild country full of laughing waters, with pines marching up alongside them circling the glassy colored lakes, full of noble windy slopes and high grassy valleys barred by the sharp, straight shadows of new mountains. All the cliffs of that country have fresh edges, and the light that cuts between them from the westering sun lies yellowly along the sod. All the winds of its open places smell of sage, and all its young rivers are swift. They begin thin and crystalline from under the forty-foot drifts, grow thick and brown in the hot leaps of early summer, run clear with full throaty laughter in midseason, froth and cloud to quick, far-off rains, fall off to low and golden-mottled rills before the first of the snows. By their changes the herder camped a hundred miles from his summer pastures knows what goes forward in them.

Let me tell you this, — every sort of life has its own zest for those who are bred to it. No more delighted sense of competency and power goes to the man who from his wire web controls the movement of money and wheat, than to the shepherd who by the passage of birds, by the stream tones, by the drift of pine pollen on the eddies of slack water, keeps tally of the pastures. Do you read the notes of mountain color as they draw into dusk? There is a color of blue, deeply pure as a trumpet tone low in the scale, that announces rain ; there is a hot blue mist suffusing into gold as it climbs against the horizon, that promises wind. There is a sense that wakes in the night with a warning to keep the flock close, and another sense of the shortest direction. The smell of the sheep is to the herder as the smack and savor of any man's work. Also it is possible to felicitate one's self on rounding a feeding flock and bringing it to a standstill within a flock-length.

The whole of that great country northward is so open and well-ordered that it affords the freest exercise of shepherd craft, every man going about to seek the preferred pastures for which use has bred a liking. Miles and miles of that district are dusky white with sage, falling off to cienàgas, — grassy hollows of seeping springs, — cooled by the windy flood that sets from the mountain about an hour before noon. The voice of that country is an open whisper, pointed at intervals by the deep whir-r-r-r of the sage hens rising from some place of hidden waters. Times when there is moonlight, watery and cold, a long thin howl detaches itself from any throat and welters on the wind. Here the lift of the sky through the palpitant, pale noons exalts the sense, and the ruffle of the sage

under it turning silverly to the wind stirs at
the heart as the slow smile of one well-loved
of whom you are yet a little afraid. Such
hours, merely at finding in the bent tops of
the brush the wattling by which the herder
keeps his head from the sun, passes the flash
and color of the time when the man-seed was
young and the Power moved toward the Par-
thenon from a plat of interlacing twigs.

The sagebrush grows up to an elevation of
eight or nine thousand feet and the wind has
not quite lapped up the long-backed drifts from
its hollows when the sheep come in. A month
later there will begin to be excellent browse
along the lower pine borders, meadow sweet,
buckthorns, and sulphur flower. The yellow
pines, beaten by the wind, or at the mere stir of
pine warblers and grosbeaks in their branches,
give out clouds of pollen dust.

The suffusion of light over the Sierra highlands is singular. Broad bands of atmosphere
infiltrating the minareted crests seem not to
be penetrated by it, but the sage, the rounded
backs of the sheep, the clicking needles of the
pines give it back in luminous particles in-

finitely divided. Airy floods of it pour about
the plats of white and purple heather and
deepen vaporously blue at the bases of the
headlands. Long shafts of it at evening fall
so obliquely as to strike far under the ragged
bellies of the sheep. Wind approaches from
the high places; even at the highest it drops
down from unimagined steeps of air. When it
moves in a cañon, before ever the near torches
of the castilleia are stirred by it, far up you hear
the crescendo tone of the fretted waters, first
as it were the foam of sound blown toward you,
and under it the pounding of the falls. Then
it runs with a patter in the quaking asp; now it
takes a fir and wrestles with it; it wakes the
brushwood with a whistle; in the soft dark of
night it tugs at the corners of the bed.

Weather warnings in a hill country are
short but unmistakable; it is not well anywhere about the Sierras to leave the camp
uncovered if one must move out of reach of it.
And if the herder tires of precautions let him
go eastward of the granite ranges where there
is no weather. Let him go by the Hot Creek
country, by Dead Man's Gulch and the Suck-

ing Sands, by the lava Flats and the pink and roan-colored hills where the lost mines are, by the black hills of pellucid glass where the sage gives place to the bitter brush, the *wheno-nabe*, where the carrion crows catch grasshoppers and the coyotes eat juniper berries, where, during the months man finds it possible to stay in them, there is no weather. Let him go, if he can stand it, where the land is naked and not ashamed, where it is always shut night or wide-open day with no interval but the pinkish violet hour of the alpen glow. There is forage enough in good years and water if you know where to look for it. Indians resorted there once to gather winter stores from the grey nut-pines that head out roundly on the eight thousand foot levels each in its clear wide space. The sand between them is strewn evenly with charred flakes of roasted cones and the stone circles about the pits are powdered still with ashes, for, as I have said, there is no weather there.

There are some pleasant places in this district, nice and trivial as the childhood re-miniscences of senility, but the great laps and folds of the cañons are like the corrugations in the faces of the indecently aged. There is a look about men who come from sojourning in that country as if the sheer nakedness of the land had somehow driven the soul back on its elemental impulses. You can imagine that one type of man exposed to it would become a mystic and another incredibly brutalized.

The devotion of the herder to the necessities of the flock is become a proverb. In a matter of urgent grazing these hairy little Bascos would feed their flocks to the rim of the world and a little over it, but I think they like best to stay where the days and nights are not all of one piece, where after the flare of the storm-trumpeting sunsets, they can snuggle to the blankets and hear the rain begin to drum on the canvas covers, and mornings see the shudder of the flock under the lift of the cloud-mist like the yellowing droves of breakers in a fog backing away from the ferries in the bay. Pleasant it is also in the high valleys where the pines begin, to happen on friendly camps of Indians come up in clans and families to gather larvæ of pine borers, *chia*, ground cherries, and sunflower seed. One could well leave the flock with the dogs for an hour to see the firelight redden on care-free faces and hear the soft laughter of the women, bubbling as hidden water in the dark.

It was not until most of the things I have been writing to you about had happened; after Narcisse Duplin had died because of Suzon Moynier, and Suzon had died; after the two Lausannes had found each other and Finot had won a fortune in a lottery and gone back to France to spend it; but not long after the wavering of the tariff and its final adjust-ment had brought the sheep business to its present status, that the flocks began to be tabooed of the natural forest lands.

One must think of the coniferous belt of the Sierra Nevadas as it appears from the top of the tremendous uplift about the head of Kern and Kings rivers, as a dark mantle laid over the range, rent sharply by the dove-grey sierra, conforming to the large contours of the moun-tains and fraying raggedly along the cañons;

a sombre cloak to the mysteries by which the drainage of this watershed is made into live rivers.

Above the pines rears a choppy and disordered surf of stone, lakes in its hollows of the clear jade that welters below the shoreward lift of waves. From the troughs of the upflung peaks the shining drifts sag back. By the time they have shortened so much that the honey flutes of the wild columbine call the bees to the upper limit of trees, the flocks have melted into the wood. They feed on the chaparral up from the stream borders and in the hanging meadows that are freed first from the flood of snow-water; the raking hoofs sink deeply in the damp, loosened soil. As the waste of the drifts gathers into runnels they follow it into filled lake basins and cut off the hope of a thousand blossomy things. Then they begin to seek out the hidden meadows, deep wells of pleasantness that the pines avoid because of wetness, soddy and good and laced by bright waters, Manache meadows girdling the red hills, Kearsarge meadows above the white-barked pines, Big meadows where the creek goes smoothly on the glacier slips, Short-Hair meadows, Tehippeti meadows under the dome where the haunted water has a sound of bells, meadows of the Twin Lakes and Middle-Fork, meadows of Yosemite, of Stinking Water, and Angustora.

Chains of meadows there are that lie along creek borders, new meadows at the foot of steep snow-shedding cliffs, shut pastures flock-journeys apart, where no streams run out and no trails lead in, and between them over the connecting moraines, over the dividing knife-blade ridges, go the pines in open order with the young hope of the forest coming up under them. No doubt meadow grasses, all plants that renew from the root, were meant for forage, and for getting at them wild grazing beasts were made fleet. But nothing other than fear puts speed in man-herded flocks. Seed-renewing plants come up between the tree boles, tufty grass, fireweed, shinleaf, and pipsisiwa; these the slow-moving flocks must crop, and unavoidably along with them the seedling pines; then as by successive croppings, forest floors are cleared, they nip the tender ends of young sap-

lings, for the business of the flock is to feed and
to keep on feeding. Where the forest intervals
afforded no more grazing, good shepherds set
them alight and looked for new pastures to
spring up in the burned districts. Who knew
how far the fire crept in the brown litter or
heard it shrieking as it ran up the tall masts of
pines, or saw the wild supplications of its pitchy
smoke? As for the shepherd, he fed forward
with the flocks over the shrubby moraines.
When the thick chaparral made difficult passage, when it tore the wool, the good shepherd
set the fire to rip out a path, and the next
year found tender, sappy browze springing
from the undying roots. The flock came to the
meadows; they fed close; then the foreplanning herder turned the creek from its course
to water it anew and the rainbow trout died
gasping on the sod.

I say the good shepherd — the man who
makes good the destiny of flocks to bear wool
and produce mutton. For what else fares he
forth with his staff and his dogs? A shepherd
is not a forester, nor is he the only sort of man
ignorant and scornful of the advantage of cov-

ered watersheds. When he first went about the
business of putting the mountain to account,
the greatest number to whom water for irrigation is the greatest good had not arrived. If
in the seventies and eighties here and there a
sheepman had arisen to declare for the Forest
Reserve, who of the Powers would have heard
him, which of the New Englanders who are
now orange-growers would have understood
his speech? In fact many did so deliver themselves. The unrestricted devotion of the pine
belt to the sheep has done us damage; but let
us say no more about it lest we be made
ashamed.

The meadow pastures make long camps and
light labors. The sheep feed out to the hill
slopes in the morning and return to the stream-
side to drink. The herder lies upon the grass,
the springy grass of the willow-skirted meadows, by the white violets of alpine meadows
where the racing waters are. Then he begins
to be busy about those curious handcrafts as
old as shepherding. He makes chain orna-
ments of horsehair, black and white, and pipe

bowls of ruddy, curled roots of manzanita. He sits with his knife and his staff of willow and covers it with interlacing patterns of carved work. There was a herder whose round was by way of Antelope Valley and Agua Hedióndo who had carved his staff from the bottom, beginning with scaly fish-tailed things through all the beasts that are and some that are not, climbing up to man. Vivian who keeps the wine-shop at Kern, Vivian the Wood Carver, had a chest in his camp with a lock of several combinations, all of hard wood, the work of his knife. But chiefly the French herder loves to spend himself on the curious keys of horn that stay the bell-leathers in the yoke, for to

the shepherd born there is no more tunable, sweet sound than the varied peal of his bells "each under each," as the flock strays in the tall chaparral. Now and then in a large flock, for distinctness, clangs the flat-toned American bell, but the best come from Gap in the Hautain Alps, and come steerage in the herder's pack, though you can buy the voiceless shell of the bell from Louis Olcese at Kern. The metal is thin and shines like the gold of Mazourka, and though it is dimmed by use like old bronze, though it colors in time as the skin of Indians, and the edge of it wears sharp as a knife-blade where it rubs along the sand, the tone of it is deep and sweet. The clapper of a French bell is a hard tip of ram's horn, or the ankle bone of a burro, hung on a soft buckskin thong, a fashion old as Araby. Shepherds from the Rhone love to stay the bells on great oak bows as broad as a man's hand, flaring at the ends; and where the bell-leathers pass through they are held by curious keys of horn. Some I have from Vivian Wright of the hard tips of bighorn, softened and shaped with infinite long care, matched perfectly for curve and color. There is a sort of fascination in the naïve and unrelated whittlings and plaitings that proceed from men who have a musing way of life, as if when the mind is a little from itself some figment of the Original Impulse begins to fumble through the teachable strong fingers toward creation. Such hints do glimmer on the sense when with his knife the herder beguiles the still noons of summer meadows.

It was there, too, I first heard the flute of the Dauphinoisa.

I had come up an hour of stiff climbing on a glacier slip, by the long shin-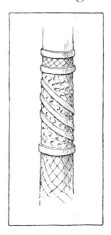ing granite bosses, treading the narrow footholds of the saxifrage, by the great plats of winy, red penstemon, odorous and hot, hugging perilously around grey, sloping, stony fronts, scarred purple by the shallow-creviced epilobium; by white-belled beds of cassiope, where a spring issued whisper-ingly on the stones; by glassy hollows of snow-water, with cool vagrant airs blowing blithely on the heather; then warm, weathered surfaces of stone with flocks of white columbine adrift about their cleavages; and above all the springy, prostrate trunks of the white-barked pine, depressed on the polished

frontage of the hill. Here I heard at intervals
the flute, sweet single notes as if the lucid air
had dripped in sound. Awhile I heard it, and
between, the slumberous roll of bells and the
whistling whisper of the pines, the long note
of the pines like falling water and water falling
like the windy tones of pines; then the warble
of the flute out of the flock-murmur as I came
over the back of the slip where it hollowed to
let in a little meadow fresh and flowered.

The herder sat with his back to a boulder
and gave forth with his breath small notes of
sweet completeness, threading the shape of a
tune as the drip of snow-water threads among
the stones, and the tune an old one such as
suits very well with a comfortable mind and a
rosy meadow. The flute was a reed, a common
reed out of Inyo, from the muddy water where
it sprawls between the marshes, and the herder
had shaped it with his knife; but it could say
as well as another that though grieving was no
doubt wholesome when grief was seasonable,
since the hour was set for gladness it was well
to be glad most completely.

VII

THE FLOCK.

CHAPTER VII

THE FLOCK

THE earliest important achievement of ovine intelligence is to know whether its own notion or another's is most worth while, and if the other's, which one. Individual sheep have certain qualities, instincts, competencies, but in the man-herded flocks these are superseded by something which I shall call the flock-mind, though I cannot say very well what it is, except that it is less than the sum of all their intelligences. This is why there have never been any notable changes in the management of flocks since the first herder girt himself with a wallet of sheepskin and went out of his cave dwelling to the pastures.

Understand that a flock is not the same thing as a number of sheep. On the stark wild headlands of the White Mountains, as many as thirty Bighorn are known to run in loose, fluctuating hordes; in fenced pastures, two to three hundred; close-herded on the range, two to three thousand; but however artificially augmented, the flock is always a conscious adjustment. As it is made up in the beginning of the season, the band is chiefly of one sort, wethers or ewes or weanling lambs (for the rams do not run with the flock except for a brief season in August); with a few flock-wise ones, trained goats, the *cabestres* of the Mexican herders, trusted bell-wethers or experienced old ewes mixed and intermeddled by the herder and the dogs, becoming invariably and finally coördinate. There are always Leaders, Middlers, and Tailers, each insisting on its own place in the order of going. Should the flock be rounded up suddenly in alarm it mills

within itself until these have come to their
own places.

If you would know something of the temper
and politics of the shepherd you meet, inquire
of him for the names of his leaders. They
should be named for his sweethearts, for the
little towns of France, for the generals of the
great Napoleon, for the presidents of Repub-
lics, — though for that matter they are all ar-
dent republicans, — for the popular heroes of
the hour. Good shepherds take the greatest
pains with their leaders, not passing them with
the first flock to slaughter, but saving them to
make wise the next.

There is much debate between herders as to
the advantage of goats over sheep as leaders.
In any case there are always a few goats in a
flock, and most American owners prefer them;
but the Frenchmen choose bell-wethers. Goats
lead naturally by reason of a quicker instinct,
forage more freely, and can find water on their
own account. But wethers, if trained with care,
learn what goats abhor, to take broken ground
sedately, to walk through the water rather than
set the whole flock leaping and scrambling;

but never to give voice to alarm as goats will, and call the herder. Wethers are more bidable once they are broken to it, but a goat is the better for a good beating. Echenique has told me that the more a goat complains under his cudgelings the surer he is of the brute's need of discipline. Goats afford another service in furnishing milk for the shepherd, and, their udders being most public, will suckle a sick lamb, a pup, or a young burro at need.

It appears that leaders understand their office, and goats particularly exhibit a jealousy of their rights to be first over the stepping-stones or to walk the teetering log-bridges at the roaring creeks. By this facile reference of the initiative to the wisest one, the shepherd is served most. The dogs learn to which of the flock to communicate orders, at which heels a bark or a bite soonest sets the flock in motion. But the flock-mind obsesses equally the best trained, flashes as instantly from the Meanest of the Flock.

Suppose the sheep to scatter widely on a heather-planted headland, the leader feeding far to windward. Comes a cougar sneaking up the trail between the rooted boulders toward the Meanest of the Flock. The smell of him, the play of light on his sleek flanks startles the unslumbering fear in the Meanest; it runs widening in the flock-mind, exploding instantly in the impulse of flight.

Danger! flashes the flock-mind, and in danger the indispensable thing is to run, not to wait until the leader sniffs the tainted wind and signals it; not for each and singly to put the occasion to the proof; but to run — of this the flock-mind apprises — and to keep on running until the impulse dies faintly as water-rings on the surface of a mantling pond. In the wild pastures flight is the only succor, and since to cry out is to interfere with that business and draw on the calamity, a flock in extremity never cries out.

Consider, then, the inadequacy of the flock-mind. A hand-fed leader may learn to call the herder vociferously, a cosset lamb in trouble come blatting to his heels, but the flock has no voice other than the deep-mouthed peal-ings hung about the leader's neck. In all that darkling lapse of time since herders began to sleep by the sheep with their weapons, affording a protection that the flock-mind never learns to invite, they have found no better trick than to be still and run foolishly. For the flock-mind moves only in the direction of the Original Intention. When at shearings or markings they run the yearlings through a gate for counting, the rate of going accelerates until the sheep pass too rapidly for numbering. Then the shepherd thrusts his staff across the opening, forcing the next sheep to jump, and the next, and the next, until, Jump! says the flock-mind. Then he withdraws the staff, and the sheep go on jumping until the impulse dies as the dying peal of the bells.

By very little the herder may turn the flock-mind to his advantage, but chiefly it works against him. Suppose on the open range the impulse to forward movement overtakes them, set in motion by some eager leaders that remember enough of what lies ahead to make them oblivious to what they pass. They press ahead. The flock draws on. The momentum of travel grows. The bells clang soft and hurriedly; the sheep forget to feed; they neglect the tender pastures ; they will not stay to drink. Under an unwise or indolent herder the sheep going on an accustomed trail will over-travel and under-feed, until in the midst of good pasture they starve upon their feet. So it is on the Long Trail you so often see the herder walking with his dogs ahead of his sheep to hold them back to feed. But if it should be new ground he must go after and press them skillfully, for the flock-mind balks chiefly at the unknown.

If a flock could be stopped as suddenly as it is set in motion, Sanger would never have lost to a single bear the five hundred sheep he told me of. They were bedded on a mesa breaking off in a precipice two hundred feet above the valley, and the bear came up behind them in the moonless watch of night. With no sound but the scurry of feet and the startled clamor of the bells, the flock broke straight ahead. The brute instinct had warned them asleep but it could not save them awake. All that the flock-mind could do was to stir them instantly to running, and they fled straight away over the headland, piling up, five hundred of them, in the gulch below.

In sudden attacks from several quarters, or inexplicable man-thwarting of their instincts, the flock-mind teaches them to turn a solid front, revolving about in the smallest compass with the lambs in the midst, narrowing and in-drawing until they perish by suffocation. So they did in the intricate defiles of Red Rock, where Carrier lost two hundred and fifty in '74, and at Poison Springs, as Narcisse Duplin told me, where he had to choose between leaving them to the deadly waters, or, prevented from the spring, made witless by thirst, to mill about until they piled up and killed threescore in their midst. By no urgency of the dogs could they be moved forward or scattered until night fell with coolness and returning sanity. Nor does the imperfect gregariousness of man always save us from ill-considered rushes or strangulous in-turnings of the social mass. Notwithstanding there are those who would have us to be flock-minded.

It is probable that the obsession of this over-sense originates in the extraordinary quickness with which the sheep makes the superior intelligence of the leader serve his

own end. A very little running in the open
range proves that one in every group of sheep
has sharper vision, quicker hearing, keener
scent; henceforth it is the business of the dull
sheep to watch that favored one. No slightest
sniff or stamp escapes him; the order for flight
finds him with muscles tense for running.

The worth of a leader in close-herded flocks
is his ability to catch readily the will of the
herder. Times I have seen the sheep feeding
far from the man, not knowing their appointed
bedding-place. The dogs lag at the herder's
heels. Now as the sun is going down the man
thrusts out his arm with a gesture that conveys
to the dogs his wish that they turn the flock
toward a certain open scarp. The dogs trot
out leisurely, circling widely to bring up the
farthest stragglers, but before they round upon
it the flock turns. It moves toward the appointed quarter and pours smoothly up the hill.
It is possible that the leaders may have learned
the language of that right arm, and in times
of quictude obey it without intervention of the
dogs. It is also conceivable that in the clear
silences of the untroubled wild the flock-mind

takes its impulse directly from the will of the herder.

Almost the only sense left untouched by man-herding is the weather sense. Scenting a change, the sheep exhibit a tendency to move to higher ground; no herder succeeds in making his flock feed in the eye of the sun. While rain falls they will not feed nor travel except in extreme desperation, but if after long falling it leaves off suddenly, night or day, the flock begins to crop. Then if the herder hears not the bells nor wakes himself by that subtle sense which in the outdoor life has time to grow, he has his day's work cut out for him in the rounding-up. A season of long rains makes short fleeces.

Summers in the mountains, sheep love to lie on the cooling banks and lick the snow, preferring it to any drink; but if falling snow overtakes them they are bewildered by it, find no food for themselves, and refuse to travel while it lies on the ground. This is the more singular, for the American wild sheep, the Bighorn, makes nothing of a twenty foot fall; in the blinding swirl of flakes shifts only to let the drifts pile under him; ruminates most content- edly when the world is full of a roaring white wind. Most beasts in bad weather drift before a storm. The faster it moves the farther go the sheep; so if there arises one of those blowy days that announce the turn of the two seasons, blinding thick with small dust, at the end of a few hours of it the shepherd sees the tails of his sheep disappearing down the wind. The tendency of sheep is to seek lower ground when disturbed by beasts, and under weather stress to work up. When any of his flock are strayed or stampeded, the herder knows by the occa- sion whether to seek them up hill or down. Seek them he must if he would have them again, for estrays have no faculty by sense or scent to work their way back to the herd. Let them be separated from it but by the roll of the land, and by accident headed in another direc- tion, it is for them as if the flock had never been. It is to provide against this incompe- tency that the shepherd makes himself markers, a black sheep, or one with a crumpled horn or an unshorn patch on the rump, easily notice-

able in the shuffle of dust-colored backs.   It
is the custom to have one marker to one hundred sheep, each known by his chosen place
in the flock which he insists upon, so that if
as many as half a dozen stray out of the band
the relative position of the markers is changed;
or if one of these conspicuous ones be missing
it will not be singly, because of the tendency
of large flocks to form smaller groups about
the best worth following.

I do not know very well what to make of
that trait of lost sheep to seek rock shelter at
the base of cliffs, for it suits with no character-
istic of his wild brethren. But if an estray in his
persistent journey up toward the high places
arrives at the foot of a tall precipice, there he
stays, seeking not to go around it, feeding out
perhaps and returning to it, but if frightened
by prowlers, huddling there to starve.   Could
it be the survival, not of a wild instinct, — it is
too foolish to have been that, — but of the cave-
dwelling time when man protected him in his
stone shelters or in pens built against the base
of a cliff, as we see the herder yet for greater
convenience build rude corrals of piled bould-

ers at the foot of an overhanging or insurmountable rocky wall? It is yet to be shown how long man halted in the period of stone dwelling and the sheep with him; but if it be assented that we have brought some traces of that life forward with us, might not also the sheep?

Where the wild strain most persists is in the bedding habits of the flock. Still they take for choice, the brow of a rising hill, turning outward toward the largest view; and never have I seen the flock all lie down at one time. Always as if by prearrangement some will stand, and upon their surrendering the watch others will rise in their places headed to sniff the tainted wind and scan the rim of the world. Like a thing palpable one sees the racial obligation pass through the bedded flock; as the tired watcher folds his knees under him and lies down, it passes like a sigh. By some mysterious selection it leaves a hundred ruminating in quietude and troubles the appointed one. One sees in the shaking of his sides a hint of struggle against the hereditary and so unnecessary instinct, but sighing he gets upon his feet. By noon or night the flock instinct never sleeps. Waking and falling asleep, waking and spying on the flock, no chance discovers the watchers failing, even though they doze upon their feet; and by nothing so much is the want of interrelation of the herder and the flock betrayed, for watching is the trained accomplishment of dogs.

The habit of nocturnal feeding is easily resumed, the sheep growing restless when the moon is full, and moving out to feed at the least encouragement. In hot seasons on the treeless range the herder takes advantage of it, making the longer siesta of the burning noon. But if the habit is to be resumed or broken off, it is best done by moving to new grounds, the association of locality being most stubborn to overcome.

Of the native instincts for finding water and knowing when food is good for them, herded goats have retained much, but sheep not a whit. In the open San Joaquin, said a good shepherd of that country, when the wind blew off the broad lake, his sheep, being thirsty, would break and run as much as a mile or two in that direction; but it seems that the alkaline

dust of the desert range must have diminished
the keenness of smell, for Sanger told me how,
on his long drive, when his sheep had come
forty miles without drink and were then so
near a water-hole that the horses scented it
and pricked up their ears, the flock became
unmanageable from thirst and broke back to
the place where they had last drunk. Great
difficulty is experienced in the desert ranges in
getting the flock to water situated obscurely
in steep ravines; they panting with water need,
but not even aware of its nearness until they
have been fairly thrust into it. Then if one lifts
up a joyous blat the dogs and the herder must
stand well forward to prevent suffocation by
piling up of the flock. You should have heard
José Jesús Lopez tell how, when the ten thousand came to water in the desert after a day or
two of dry travel, when the first of the nearing
band had drunk he lifted up the water call;
how it was taken up and carried back across
the shouldering brutes to the nearest band behind, and by them flatly trumpeted to the next,
and so across the mesa, miles and miles in the
still, slant light.

When Watterson ran his sheep on the plains
he watered them at a pump, and in the course
of the season all the bands that bore the Three
Legs of Man got to know the smell pertaining
to that brand, drinking at the troughs as they
drew in at sundown from the feeding-ground.
But when for a price strange bands in passing
drank there, he could in no wise prevail upon
his own sheep to drink of the water they had
left. The flocks shuffled in and sniffed at the
tainted drink and went and lay down waterless.
The second band drew alongside and made as
if to refresh themselves at the troughs, but
before they had so much as smelled of it: —

Ba-a-a, Ba-a-a-a! blatted the first flock, and
the newcomers turned toward them and lay
down. Comes another band and the second
takes up the report, not having proved the
event but accepting it at hearsay from the
first.

Ba-a-a-a-d, Ba-a-a-a-d! blat the watchers, and
when that has happened two or three times
the shepherd gives over trying to make his
sheep accept the leavings of the troughs, whatever the price of water, but turns it out upon

the sand. Sheep will die rather than drink water which does not please them, and die drinking water with which they should not be pleased. Nor can they discriminate in the matter of poisonous herbs. In the northerly Sierras they perish yearly, cropping the azaleas; Julien lost three or four hundred when wild tobacco (*nicotiana attenuata*) sprang up after a season of flood water below Coyote Holes; and in places about the high mountains there are certain isolated meadows wherein some herb unidentified by sheepmen works disaster to the ignorant or too confiding herder. Such places come to be known as Poison Meadows, and grasses ripen in them uncropped year after year. Yet it would seem there is a rag-tag of instinct left, for in the desert regions where sheep have had a taste of Loco-weed (*astragalus*) which affects them as cocaine, like the devotees of that drug, they return to seek for it and become dopy and worthless  through its excess; and a flock that has suf-

fered from milkweed poisoning learns at last
to be a little aware of it. Old tales of folk-
lore would have us to understand that this
atrophy of a vital sense is within the reach of
history. Is it not told indeed, in Araby, that
the exhilaration of coffee was discovered by a
goatherd from the behavior of his goats when
they had cropped the berries ?

By much the same cry that apprises the flock
of tainted drink they are made aware of strangers in the band. This is chiefly the business
of yearlings, wise old ewes and seasoned wethers not much regarding it. One of the band
discerns a smell not the smell of his flock, and
bells the others to come on and inquire. They
run blatting to his call and form a ring about
the stranger, vociferating disapproval until the
flock-mind wakes and pricks them to butt the
intruder from the herd; but he persisting and
hanging on the outskirts of the flock, acquaints
them with his smell and becomes finally incor-
porate in the band. Nothing else but the rat-
tlesnake extracts this note of protest from the
flock. Him also they inclose in the noisy ring
until the rattler wriggles to his hole, or the

herder comes with his *makila* and puts an end
to the commotion.

It is well to keep in mind that ordinarily
when the flock cries there is nothing in particular the matter with it. The continuous
blether of the evening round-up is merely the
note of domesticity, ewes calling to their lambs,
wethers to their companions as they revolve to
their accustomed places, all a little resentful of
the importunity of the dogs. In sickness and
alarm the sheep are distressfully still, only
milkweed poisoning, of all evils, forcing from
them a kind of breathy moan; but this is merely
a symptom of the disorder and not directed
toward the procurement of relief.

It is doubtful if the herder is anything more
to the flock than an incident of the range,

except as a giver of salt, for the only cry they make to him is the salt cry. When the natural craving is at the point of urgency they circle about his camp or his cabin, leaving off feeding for that business; and nothing else offering, they will continue this headlong circling about a boulder or any object bulking large in their immediate neighborhood remotely resembling the appurtenances of man, as if they had learned nothing since they were free to find licks for themselves, except that salt comes by bestowal and in conjunction with the vaguely indeterminate lumps of matter that associate with man. As if in fifty centuries of man-herding they had made but one step out of the terrible isolation of brute species, an isolation impenetrable except by fear to every other brute, but now admitting the fact without knowledge, of the God of the Salt. Accustomed to receiving this miracle on open boulders, when the craving is strong upon them they seek such as these to run about, vociferating, as if they said, In such a place our God has been wont to bless us, come now let us greatly entreat Him. This one quavering bleat, unmistakable to the sheepman even at a distance, is the only new note in the sheep's vocabulary, and the only one which passes with intention from himself to man. As for the call of distress which a leader raised by hand may make to his master, it is not new, is not common to flock usage, and is swamped utterly in the obsession of the flock-mind.

But when you hear shepherds from the Pyrenees speak of the salt call it is no blether of the sheep they mean, but that long, rolling, high and raucous *Ru-u-u-u-u-u* by which they summon the flock to the lick. And this is most curious that no other word than this is recognized as exclusive to the sheep, as we understand " scat " to be the peculiar shibboleth of cats, and " bossy " the only proper appellate of cows. Ordinarily the herder does not wish to call the sheep, he prefers to send the dogs, but if he needs must name them he cries Sheep, sheep! or *mouton*, or *boregíto*, as his tongue is, or apprises them of the distribution of salt by beating on a pan. Only the Basco, and such French as have learned it from him, troubles his throat with this searching, mutilated cry. If

it should be in crossing the Reserve when the
rangers hurry him, or on the range when in
the midst of security, suddenly he discovers the
deadly milkweed growing all abroad, or if above
the timber-line one of the quick, downpouring
storms begins to shape in the pure aerial
glooms, at once you see the herder striding at
the head of his flock drawing them on with the
uplifted, *Ru-u-u-uuuuu !* and all the sheep
running to it as it were the Pied Piper come
again.

Suppose it were true what we have read, that
there was once an Atlantis stationed toward
the west, continuing the empurpled Pyrenees.
Suppose the first of these Pyrenean folk were,
as it is written, just Atlantean shepherds straying farthest from that happy island, when the
seas engulfed it; suppose they should have carried forward with the inbred shepherd habit
some roots of speech, likeliest to have been
such as belonged to shepherding — well then,
when above the range of trees, when the wild
scarps lift rosily through the ineffably pure blue
of the twilight earth, suffused with splendor of
the alpen glow, when the flock crops the tufted

grass scattering widely on the steep, should you
see these little men of long arms leaping among
the rocks and all the flock lift up their heads
to hear the ululating *Ru-u-ubru-u-uuu!* would
not all these things leap together in your mind
and seem to mean something? Just suppose!

VIII

THE GO–BETWEENS — A
CHAPTER TO BE OMITTED BY
THE READER WHO HAS NOT
LOVED A DOG.

CHAPTER VIII

THE GO-BETWEENS

WHAT one wishes to know is just what the
dog means to the flock. It might be something
of what the dark means to man, the mould of
fear, the racial memory of the shape in which
Terror first beset them. It is as easy to see
what the flock means to the dog as to understand what it meant before man went about this
business of perverting the Original Intention.
If it is a trick man has played upon the dog to
constitute him the guardian of his natural
prey, he has also been played upon, for even

as men proved their God on the persons of
the brethren and exterminated tribes to show
how great He was, latterly they afflict themselves to offer up the heathen scathless and
comforted.

Now that in the room of the Primal Impulse,
the herder is the god of the sheep dog, the
flock is become an oblation. The ministrant
waits with pricked ears and an expectant eye
the motion of his deity; he invites orders by
eagerness; he worries the sheep by the zeal-
ousness of care; that not one may escape he
threads every wandering scent and trails it
back to the flock. In short, when in the best
temper for his work he frequently becomes useless from excess of use. But in the half a hundred centuries that have gone to perverting his
native instincts, the sheep have hardly come
so far. They no longer flee the herd dog, but
neither do they run to him. When he rounds
them they turn; when he speaks they tremble;
when he snaps they leave off feeding; but when
they hear his cousin-german, the coyote, pad-
ding about them in the dark, they trust only to
fleeing. For this is the apotheosis of the dog,

that he fights his own kind for the flock, but the flock does not know it.

It is notable that the best sheep dogs are most like wolves in habit, the erect triangular ears, the long thin muzzle, the sag of the bushy tail, the thick mane-like hackles; as if it were on the particular aptness for knowing the ways of flocking beasts developed by successful wolves that the effective collie is moulded. No particular breed of dogs is favored by the herders hereabout, though Scotch strains predominate. Among the Frenchmen a small short-tailed, black-and-white type is seen oftenest, a pinto with white about the eyes. One may pay as much as five dollars or five hundred for a six months' pup, but mostly the herders breed their own stock and exchange among themselves. Ordinarily the dog goes with the flock, is the property of the owner, for sheep learn to know their own guardian and suffer an accession of timidity if a stranger is set over them.

The herder who brings up a dog by hand loves it surpassingly. There was one of my acquaintance had so great an attachment for a bitch called Jehane that he worked long for a hard master and yearly tendered him the full of his wage if only he might have Jehane and depart with her to a better employment. He was not single in his belief that Jehane regarded him with a like affection, for the faith a herder grows to have in the dog's understanding is only exceeded by the miracle of communication. To see three or four shepherds met in a district of good pastures, leaning on their staves, each with a dog at his knees quick and attentive to the talk, is to go a long way toward conviction.

Many years ago, but not so long that he can recall it without sorrow, Giraud lost a dog on Kern River. There had come one of the sudden storms of that district, white blasts of hail and a nipping wind; it was important to get the sheep speedily to lower ground. The dog was ailing and fell behind somewhere in the white swarm of the snow. When it lay soft and quiet over all that region and the flock was bedded far below it in the cañon, Giraud returned to the upper river, seeking and calling; twenty days he quested bootless about the

meadows and among the cold camps. More he
could not have done for a brother, for Pierre
Giraud was not then the owner of good acres
and well-fleeced Merinos that he is now, and
twenty days of a shepherd's time is more than
the price of a dog. " And still," Pierre finishes
his story simply, " whenever I go by that country of Kern River I think of my dog."

Curiously, the obligation of his work — who
shall say it is not that higher form of habit out
of which the sense of duty shapes itself ? — is
always stronger in the dog than the love of the
herder. Lacking a direct command, in any
severance of their interests, the collie stays by
the sheep. In that same country of young roaring rivers a shepherd died suddenly in his
camp and was not found for two days. The

flock was gone on from the meadow where he lay, straying toward high places as shepherd-less sheep will, and the dogs with them. They had returned to lick the dead face of the herder, no doubt they had mourned above him in their fashion in the dusk of pines, but though they could win no authority from him they stayed by the flock. So they did when the two herds-men of Barret's were frozen on their feet while still faithfully rounding the sheep; they dropped stilly in their places and were over-blown by the snow. The dogs had scraped the drifts from their bodies, and the sheep had trampled mindlessly on the straightened forms, but at the end of the third day when succor found them, the dogs had come a flock-journey from that place and had turned the sheep toward home. This is as long as can be proved that the sense of responsibility to the flock stays with the dog when he feels himself aban-doned by his over-lord.

A dog might remain indefinitely with the sheep because he has the habit of association, but the service of herding is rendered only at the bidding of the gods. The superintendent

of Tejon told me of a dog that could be trusted
to take a bunch of muttons that had been cut
out for use at the ranch house, and from any
point on the range, drive them a whole day's
journey at his order, and bring them safely to
the home corral. Señor Lopez, I think, related of another that it was sent out to hunt
estrays, and not returning, was hunted for and
found warding a ewe and twin lambs, licking
his wounds and sniffing, not without the appearance of satisfaction, at a newly killed coyote. The dog must have found the ewe in
travail, for the lambs were but a few hours old,
and been made aware of it by what absolute
and elemental means who shall say, and stood
guarding the event through the night.

At Los Alisos there was a bitch of such excellent temper that she was thought of more
value for raising pups than herding; she was,
therefore, when her litter came, taken from the
flock and given quarters at the ranch house.
But in the morning Flora went out to the sheep.
She sought them in the pastures where they
had been, and kept the accustomed round, returning wearied to her young at noon; she fol-

lowed after them at evening and covered with panting sides the distance they had put between them and her litter. At the end of the second day when she came to her bed, half dead with running, she was tied, but gnawed the rope, and in twenty-four hours was out on the cold trail of the flock. One of the vaqueros found her twenty miles from home, working faint and frenzied over its vanishing scent. It was only after this fruitless sally that she was reconciled to her new estate.

Now consider that we have very many high and brave phrases for such performances when they pertain to two-footed beings who grow hair on their heads only, and are disallowed the use of them for the four-foots that have hair all over them. Duty, chivalry, sacrifice, these are words sacred to the man things. But how shall one loving definiteness consign to the loose limbo of instinct all the qualities engendered in the intelligence of the dog by the mind of man? For it is incontrovertible that a good sheep dog is made.

The propensity to herd is fixed in the breed. Some unaccountably in any litter will have missed the possibility of being good at it, and a collie that is not good for a herd dog is good for nothing. The only thing to do with the born incompetent is to shoot it or give it to the children ; in the bringing up of a family almost any dog is better than no dog at all. What good breeding means in a young collie is not that he is fit to herd sheep, but that he is fit to be trained to it. Aptitude he may be born with, but can in no wise dispense with the hand of the herder over him. What we need is a new vocabulary for the larger estate which a dog takes on when he is tamed by a man.

Training here is not carried to so fine a pitch as abroad, most owners not desiring too dependable a dog. The herder is the more likely to leave the flock too much to his care, and whatever a sheep dog may learn, it is never to discriminate in the matter of pasture. An excellent collie makes an indolent herder.

Every man who follows after sheep will tell you how he thinks he trains his pups, and of all the means variously expounded there are two that are constant. It is important that the dog acquire early the habit of association, and to this purpose herders will often carry a pup in the cayaca and suckle it to a goat. Most important is it that he shall learn to return of his own motion to the master for deserved chastisement. To accomplish this the dog is tied with sufficient ropeway and punished until he discovers that the ease of his distress is to come straightly to the hand that afflicts him. He is to be tied long to allow him room for volition and tied securely that he may not once get clean away from the trainer's hand. Once a dog, through fear or the sense of anger incurred, escapes his master for a space of hours, there is not much to be done by way of retrievement. It is as if the impalpable bridge between his mind and the mind of man, being broken by the act, is never to be built again. For this in fine is what constitutes a good herd dog, to be wholly open to the suggestion of the man-mind, and carry its will to the flock. His is the service of the Go-Between. Not that he knows or cares what becomes of the flock, but merely what the herder intends toward it.

I have said the shepherd will tell you how he thinks he trains his collies, for watching them I grow certain that more goes forward than the herder is rightly aware. Working communication between them is largely by signs, since the dog manœuvres at the distance of a flock-length, taking orders from the herder's arm. Every movement of the flock can be so effected, but if the herder would have barking, he must say to him, Speak, and he speaks. The teaching methods seem not to be contrived by any rule, as if every man fumbling at the dog's understanding had hit upon a device which seemed to accomplish his end, and might or might not serve the next adventure. You would not suppose in any other case that by waving arms, buffets, pettings, and retrievings, and by no other means, so much could be communicable in violation to racial instincts, with no root in experience and only a possible one in the generational memory; nor do I for one suppose it. Moreover it sticks in my mind that I have never seen one herd dog instruct another even by the implication of behaving in such a manner as to invite imitation.

Bobcats I have seen teaching their kittens to seek prey, young eagles coached at flying, coyote cubs remanded to the trail with a snarl when wishful to leave it; but never the sheep dog teaching her young to round and guard. In this all the shepherds of the Long Trail bear me out. Assuredly the least intelligent dog learns something by imitation; to be convinced of it one has only to note the assumed postures, the look as of a very deaf person who wishes to have you believe that he has heard, the self-gratulation when some tentative motion proves acceptable, the tolerable assumption when it fails that the sally has been undertaken merely by way of entertainment. But with it all no intention of being imitated.

Since all these things are so, how then can a shepherd say to the Go-Between what the dog cannot say to another dog? It is not altogether that they lack speech, for, as I say, the work of herding goes on by signs, and I have come to an excellent understanding with some collies that know only Basque and a patois that is not the French of the books. Fellowship is helped by conversation, though it is not indispensable, and if the herder has an arm to wave has not the dog a tail to wag? If he reads the face of his master, and who that has been loved by a dog but believes him amenable to a smile or a frown, may he not so learn the countenance of his blood brother? Notwithstanding, the desire of the shepherd which the dog bears to the sheep remains with respect to other dogs, like the personal revelation of a deity, locked, incommunicable. He arises to the man virtues so long as the man's command, or the echo of it, lies in his consciousness. But we, when we have arrived at the pitch of conserving what was once our study to destroy, conceive that we have done it of ourselves.

What a herd dog has first to learn is to know every one of two to three hundred sheep, and to know them both by sight and smell. This he does thoroughly. When Watterson was running sheep on the plains he had a young collie not yet put to the herd but kept about the pumping plant. As the sheep came in by hundreds to the troughs, the dog grew so to know them that when they had picked up an estray from another band he discovered it from afar off, and darting as a hornet, nipping and yelping, parted it out from the band. At that time no mere man would have pretended without the aid of the brand to recognize any of the thousands that bore it.

How long recollection stays by the dog is not certain, but at least a twelvemonth, as was proved to Filon Gerard after he had lost a third of his band when the Santa Anna came roaring up by Lone Pine with a cloud of saffron-colored dust on its wings. After shearing of next year, passing close to another band, Filon's dogs set themselves unbidden to routing out of it, and rounding with their own, nearly twenty head which the herder, being an honest man, freely admitted he had picked up on the mesa following after Filon the spring before.

Quick to know the willful and unbidable members of a flock, the wise collie is not sparing of bites, and following after a stubborn estray will often throw it, and stand guard until help arrives, or the sheep shows a better mind. But the herder who has a dog trained at the difficult work of herding range sheep through

the chutes and runways into boats and cars for
transportation is the fortunate fellow.

There was Pete's dog, Bourdaloue, that, at
the Stockton landing, with no assistance, put
eight hundred wild sheep from the highlands
on the boat in eight minutes, by running along
the backs of the flock until he had picked out
the stubborn or stupid leaders that caused the
sheep to jam in the runway, and by sharp bites
set them forward, himself treading the backs
of the racing flock, like the premier equestri-
enne of the circus, which all the men of the
shipping cheered to see.

In shaping his work to the land he moves in,
an old wolf-habit
of the sheep dog
comes into play.
From knowing how
to leap up in mid-
run to keep sight
of small quarry, the
collie has learned to
mount on stumps
and boulders to ob-

serve the flock. So he does in the sage and

chamisal, and of greater necessity years ago in
the coast ranges where the mustard engulfed
the flock until their whereabouts could be
known only by the swaying of its bloom.
Julien, the good shepherd of Lone Pine, had
a little dog, much loved, that would come and
bark to be taken up on his master's shoulder
that he might better judge how his work lay.
The propensity of sheep to fall over one
another into a pit whenever occasion offers
is as well noted by the dog as the owner; so
that there was once a collie of Hittell's of such
flock-wisdom that at a point in a certain drive
where an accident had occurred by the sheep
being gulched, he never failed afterward to
go forward and guard the bank until the flock
had gone by.

Footsoreness is the worst evil of the Long
Trail; cactus thorn, foxtail, and sharp, hot
granite sands induce so great distress that to
remedy it the shepherd makes moccasins of deer-
skin for his dogs. Once having experience of
these comforts the collie returns to the herder's
knee and lifts up his paws as a gentle invitation to have them on when the trail begins to

wear. On his long drive Sanger had slung a
rawhide under the wagon to carry brushwood
for the fire, but the dogs soon discovered in it
a material easement of their fatigues, and
would lie in it while the team went forward,
each collie rousting out his confrère and insisting on his turn.

When one falls in with a sheep camp it is
always well to inquire concerning the dogs; the
herder who will not talk of anything else will
talk of these. You bend back the springy
sage to sit upon, the shepherd sits on a brown
boulder with his staff between his knees, the
dogs at his feet, ears pointed with attention.
He unfolds his cigarette papers and fumbles
for the sack.

" Eh, my tobacco? I have left it at the camp;
go, Pinto, and fetch it."

Away races the collie, pleased as a patted
schoolboy, and comes back with the tobacco
between his jaws.

" I must tell you a story of that misbegotten
devil of a he goat, Noé," says the shepherd,
rolling a cigarette; " you, go and fetch Noé that
Madame-who-writes-the-book may see."

In a jiffy the dog has nipped Noé by the
ankles and cut him out of the band, but you
will have to ask again before you get your story,
for it is not Noé the shepherd has in mind. In
reality he is bursting with pride of his dog, and
thinks only to exhibit him.

It is the expansiveness of affection that ele-
vates the customary performance to an achievement. As for the other man's dog, why should
it not do well? unless his master being a dull
fellow has spent his pains to no end. But in
the Pinto there with the listening ears and
muzzle delicately pointed and inquiring, with
the eye confident and restrained as expressing
the suspension of communication rather than
its incompleteness, you perceive at once a tan-
gible and exceptionable distinction.

IX

THE STRIFE OF THE HERDSMEN — HOW THE GREAT GAME IS PLAYED IN THE FREE PASTURES, AND THE CATTLEMEN'S WAR.

CHAPTER IX

THE STRIFE OF THE HERDSMEN

THE mesa was blue with the little blue larkspur the Indians love; a larkspur sky began somewhere infinitely beyond the Sierra wall and stretched far and faintly over Shoshone Land. The ring of the horizon was as blue as the smoke of the deputy sheriff's cigar as he lay in the shade of a boulder and guessed almost by the manner of the dust how many and what brands stirred up the visible warning of their approach. The spring passage of the flocks had begun, and we were out after the tax.

Two banners of dust went up in the gaps of
the Alabamas and one below the point, two at
Symmes Creek, one crowded up under Williamson, one by the new line of willows below
Piñon, that by the time the shadows of the
mountains had shrunk into their crevices,
proved by the sound of the bells to be the flock
of Narcisse Duplin. The bell of Narcisse's
best leader, Le Petit Corporal, was notable;
large as a goat-skin wine-bottle, narrowing at
the mouth, and so long that it scraped the sand
when the Corporal browsed on the bitter brush
and lay quite along the ground when he cropped
the grass. The sound of it struck deeply under
all the notes of the day, and carried as far as
the noise of the water pouring into the pot-hole
below Kearsarge Mill.

The deputy sheriff had finished his cigar,
and begun telling me about Manuel de Borba
after he had killed Mariana in the open below
Olancha. Naylor and Robinson bought the
flock of him in good faith, though suspicion
began to grow in them as they came north
with it toward the place where Mariana lived;
then it spread in Lone Pine until it became

a rumor and finally a conviction. Then Relles Carrasco took up the back trail and found, at the end of it, Mariana lying out in the sage, full of knife wounds, and the wounds were in his back. When the deputy had proceeded as far as the search for de Borba, Narcisse came up with us.

Where we sat the wash of Pine Creek was shallow, and below lay the rude, tottering bridge of sticks and stones, such as sheepmen build everywhere in the Sierras for getting sheep across troublesome streams. Here in the course of the day came all the flocks we sighted, with others drifting into view in the south, and at twilight tide a dozen of their fires blossomed under Kearsarge in the dusk. The sheriff counted the sheep as they went singly over the bridge, with his eyes half shut against the sun and his finger wagging; as for me, I went up and down among the larkspur flowers, among the lupines and the shining bubbles of mariposa floating along the tops of the scrub, and renewed acquaintance.

" Tell me," I said to Narcisse, who because of the tawny red of his hair, the fiery red of his face, the russet red of his beard, and the red spark of his eye, was called Narcisse the Red, "tell me what is the worst of shepherding?"

"The worst, madame, is the feed, because there is not enough of it."

"And what, in your thinking, is the best?"

"The feed, madame, for there is not enough of it."

"But how could that be, both best and worst?"

Narcisse laughed full and throatily, throwing up his chin from the burned red chest all open to the sun. It was that laugh of Narcisse's that betrayed him the night he carried away Suzon Moynier from her father's house.

"It is the worst," said he, "because it is a great distress to see the flock go hungry, also it is a loss to the owner. It is the best, because every man must set his wits against every other. When he comes out of the hills with a fat flock and good fleeces it is that he has proved himself the better man. He knows the country better and has the greater skill to keep other men from his pastures. How else but by contriving shall a man get the feed

from the free pastures when it goes every year
to the best contriver? You think you would
not do it? Suppose now you have come with
a lean flock to good ground sufficient for yours
only, and before the sheep have had a fill of it,
comes another blatting band working against
the wind. You walk to and fro behind your
flock, you take out a newspaper to read, you
unfold it. Suddenly the wind takes it from
your hand, carries it rustling white and fear-
some in the faces of the approaching flock.
Ah, bah! Who would have supposed they
would stampede for so slight a thing? And by
the time their herder has rounded them up,
your sheep will have all the feed."

When Narcisse Duplin tells me this the
eyes of all the herders twinkle; glints of amuse-
ment run from one to another like white hints
of motion in the water below the birches.

"It is so," said Octavieu, the blue-eyed
Basque, "the feed is his who can keep it.
Madame goes much about the Sierras, have
you not seen the false monuments?"

"And been misled by them."

"They were not meant for such as madame,

but one shepherd when he finds a good meadow makes a false trail leading around and away from it, and another shepherd coming is deceived thereby, and the meadow is kept secret for the finder."

When Octavieu tells me this I recall a story I have heard of Little Pete, how when he had turned his flocks into an upper meadow he met a herder bound to that same feeding-ground, and by a shorter route; but the day saved him. No matter how much they neglect the calendar, French shepherds always know when it is the fourteenth of July, as if they had a sense for divining it much as gophers know when *taboose* is good to eat. Pete dug up a bottle from his cayaques.

" *Allons, mon vieux, c'est le quatorze Juillet,*"

 cried the strategist; " come, a toast; *Le Quatorze Juillet !* "

" *Le Quatorze Juillet !* "

The red liquor gurgled in their throats. Never yet was a Frenchman proof against patriotism and wine and good company. The arrested flock shuffled and sighed while Pete and their master through the rosy glow of wine saw the Bastile come down and the Tricolor go up. Incidentally they saw also the bottom of the bottle, and by that time Pete's flock was in full possession of the meadow. Pete laughs at this story and denies it, but so light-heartedly that I am sure that if it never happened it was because he happened never to think of it.

" However, I will tell you a true story," said he. " I was once in a country where there was a meadow with springs and much good feed in that neighborhood, but unwatered, so that if a man had not the use of the meadow he could get no good of it. The place where the spring was, being patented land, belonged to a man whose name does not come into the story. I write to that man and make him a price for the water and the feed, but the answer is not come. Still I think sure to have it, and leave word that the letter is to be sent to me at the camp, and move my flock every day toward the meadow. Also I observe another sheepman feeding about my trail, and I wish greatly for that letter, for I think he makes the eyes at that pasture with springs.

"All this would be no matter if I could trust my herder, but I have seen him sit by the other man's fire, and I know that he has what you call the grudge against me. For what? How should I know? Maybe there is not garlic enough in camp, maybe I keep the wine too close; and it is written in the foreheads of some men that they should be false to their employers. When it is the better part of a week gone I am sure that my herder has told the other man that I have not yet rented the springs, so I resolve at night in my blanket what I shall do. That day I send out my man with his part of the sheep very far, then I write me a letter, to me, Pierre Giraud, and put it in the camp. It is stamped, and altogether such as if it had come from the Post Office. Then I ride about my business for the day, and at night when I come late to the camp there is the herder who sings out to me and says:—

"'Here is your letter come.'"

Pete chuckles inwardly with true artistic

appreciation of finesse. " Eh, if you do this sort of thing it should be done thoroughly. I see the herder watch me with the tail of his eye while I make to read the letter.

" ' Is it right about the meadow? ' says he.

" ' You can see,' say I, and I hand him the paper, which he cannot read, but he will not confess to that. That night he goes to the other man's fire, and the next day I see that that one drops off from my trail, and I know he has had word of my letter. Then I move my sheep up to the meadow of springs."

" And the real letter, when it came — if it came ? "

" That you should ask me ! " cries Pete, and I am not sure if I am the more convinced by the reproachful waggings of his head or the deep, delighted twinkle of his eye.

In the flanking ranges east from the Sierras are few and far between water-holes the possession of which dominates great acreage of tolerable feed. For the control of them the herders strive together as the servants of Abraham and Abimelech for the wells which Abraham digged.

There was a herder once out of Dauphiny who went toward Panamint and found a spring of sweet water in a secret place. The pasture of that country was bunch grass and mesquite, and the water welled up from under the lava rock and went about the meadow to water it. When he had fed there for a fortnight and there was still grazing in the neighborhood for a month more, he looked out across the mesquite dunes and saw the dust of a flock. Then he considered and took a pail and went a long way out to meet it. Where the trail of the sheep turned into the place of the secret spring, but more than a mile from it, there was also a pool of seepage water, but muddied and trampled by the sheep. When he had come to this place the shepherd scooped out a hollow and made believe to dip up the water where it ran defiled into the hole he had digged, while the stranger came on with his flock.

It seems that at shearings and lambing where they met they were very good friends, but on the range —

"How goes the feed, *mon vieux?*"

"Excellently well, *mon ami.*"

" And the water? "

" Ah, you can see." The herder cast a contemplative eye at the turgid liquid in the pail; assuredly no sheep would drink of it. Also he looked at the feed and sighed, for it was good feed, but one really must have water.

" I think of moving to-day," said the first shepherd, but the second drew off his flock at once and returned by another trail.

The desire to be beforehand with the feed becomes an obsession; herders of the same owner will crowd each other off the range. The Manxman told me that once he had a head shepherd who played the flocks in his charge one against another, like a man cheating himself at solitaire. Though there grows tacitly among the better class of sheepmen the understanding that long-continued use establishes a sort of priority in the pastures, among themselves the herders will still be " hogging the feed."

When Sanger went on his little exodus to Montana, he went out by way of Deep Springs Valley to cross Nevada, that same valley where Harry Quinn, hoping for winter pastures in '74, lost all but twenty-two hundred out of a flock of twenty-two thousand in the only deep snow that fell there, drifting over the low, stubby shrubs shoulder high to the sheep. When Sanger first broke trail across it there was feed enough, more than enough, if pastured fairly; but out of Deep Springs came another shepherd, taking the same general direction, but forging always ahead, forcing his flock out by dawn light to get the top of the grazing. Sanger considered and made sure of the other man's intention. Presently they came to a pleasant place of springs.

"Now," said Sanger, hiding his purpose behind the honestest blue eyes and an open German countenance, "the feed is good and I can rest here some days." So assured, the enemy slept with his flock and woke late to see the dust of Sanger's sheep, kept moving in the night, vanishing northward on his horizon. And Sanger is not the only man who has been sharpened to the business by being first a settler in the time when every season called for some new contrivance against the herder's plan of feeding out the homesteader; though when he became a sheepman it is doubtful if he could have been drawn off from pasture by his own device of sprinkling salt on the range in the face of the herders so that they turned their flocks away from that country in great alarm, reporting the feed to be poisoned, a reprisal not uncommon in the early sixties.

It is also allowable, finding intruders on your accustomed ground, to burn their corrals and destroy their bridges. Meaner measures than this are not often resorted to, though there are instances.

One of the guardians of the flock whose brand is the Three Legs of Man, working up a shallow cañon toward the summer meadows, found a pertinacious Portuguese herder feeding in that direction. The flocks of the Manxman had the advantage of the near side of the cañon, and all the clear afternoon they manœuvred forth and back to keep in front of the Portuguese, he drawing close until the commingling dust of their bands hid all his motions in a golden blurr. They looked for him to break through at this point, or for some mischief which should stampede the flock, but nothing other than the quickened scurry of feet and the jangle of the bells came out of the thick haze of dust. When it cleared, the enemy was shown to have turned off sharply in retreat. The rate of his going, as well as the unexpectedness of it, bred suspicion. Not, however, until the Manxman rounded up did he discover that the fellow had, under cover of the dust, incorporated with his own band and carried away a bunch of best merinos.

Recovery of stolen sheep, detected in time, is not difficult; a much harder matter for the shepherd to explain how sheep not of his brand came in his keeping. If he is sensible he does not try to do so, and if they have come legitimately as being gathered up after a storm, accepts a small sum for their care and restores them to the claimant. If, however, they have been passed to an accomplice and out of the country, rebranded and marked anew, there is little to be done about it. For the most part, all the business amenities prevail on the open range, for this also is a part of the Great Game.

Every quarter section of land in the neighborhood of a watershed is potentially irrigable and attracts settlement. We breed yearly enough men of such large hopefulness as to be willing to live on that possibility, or of an incurable inability to live anywhere else. Ordinarily they put more zest into the struggle for the use of grazing lands that they do not own than improving those they do, but here in California there has not been between these and the cattlemen the bitterness and violence that grow out of the struggle for the range in Montana and Arizona. But for the sake of what I shall have to say touching the matter of the Forest Reserves, I shall put the case to you as it is handed up to me by men whose business has been much about the open range. In this it is well to be explicit though I appear as a mere recorder.

Two years out of three there is not pasture enough for the whole number of flocks and herds to grow fat. In good seasons they feed in the same district without interference, but sheep are close croppers, and in excessive dry years cut off the hope of renewal by eating into the root-stocks of the creeping grasses.

Their droppings also are an offense, and being herded in a bunch they defile the whole ground. After rains the grass springs afresh and the scent passes into the earth, but in the rainless Southwest it lies long and renders objectionable the scanty grass. Set against this that cattle perform the same office of fouling the pastures, so that even in starvation times one notes the flock veering away from the fresh rings of grass where cattle have passed; also the horned cattle love oozy standing ground, and work even their own distress by trampling out the springs. In the Southwest where the land is not able to bear them because of their numbers and the sheep get advantage by reason of their close method of herding, the cattlemen retort with violence. They charge the flock and run it over a cliff, or breaking into the corrals engage in disgusting butchery the like of which has not yet been imputed to herders. Also there have been killings of men, herders dropped stilly in the middle of the flock, cowboys crumpling forward in the saddle at the crossing of the trails.

The mutual offenses being as I have set

them forth, it is to be seen that much is to be
imputed to mere greed and the desire for mas-
tery. Moreover it is indisputably allowed by
cowmen that they are inherently, and on all
occasions, better than any sheepman that ever
lived. I being of neither party will not sub-
scribe to it, for the seed of that ferment which
makes caste between classes of men, the sums
of whose intelligence and right dealing are not
appreciably different, is not in me.

Just at this point it is well to recall that of
all the men who grow rich by hides and fleeces,
not one in ten does so on his own land. All
these millions of acres of mesquite and sage
and herd grass and alfilaria belong to Us.
Supinely we let them out to be the prize of
trickery and violence. That is why there can

be so few reprisals at law for offenses done on the range. What is no man's no man can be remanded for taking strongly. Consider then the simplicity of allotting fixed pastures of public lands by rental. But the present arrangement is our superior way of being flock-minded.

X

 LIERS–IN–WAIT — WHAT THEY DO TO THE FLOCK, AND WHAT THE SHEPHERD DOES TO THEM.

CHAPTER X

THERE is a writer of most agreeable animal stories who takes pains modestly to disclaim any participation in the event, but in fact he need hardly be at so much trouble. It is not the man to whom such adventures occur as by right who makes a pretty tale of them, and I am oftenest convinced of the truth of an incident in an ancient piece of writing rather misdoubted these wordy days, because it is so much in the manner of people to whom these things happen in their way of life. It is also an ex-

cellent model for an animal story and is told
in three sentences :

"Then went Samson down . . . to Timnath
. . . and behold a young lion roared against
him. . . . And he rent him as he would have
rent a kid . . . but he told not his father or
his mother what he had done."

"Jean Baptiste," say I, "where did you get
that splendid lynx skin in your cayaca?"

"Eh, it was below Olancha about moonrise
that he sprung on the fattest of my lambs. I
gave him a crack with my staff, and the dogs
did the rest."

You will hardly get a more prolix account
from any herder, though there are enough of
these tufted lynxes about the dry washes to
make their pelts no uncommon plunder of the
camps. It is only against man contrivances,
such as a wool tariff or a new ruling of the
Forestry Bureau, that the herder becomes
loquacious. Wildcats, cougars, coyotes, and
bears are merely incidents of the day's work,
like putting on stiff boots of a cold morning,
running out of garlic, or having the ewes cast
their lambs. As for weather stress, they endure

it much in the fashion of their own sheep, which if they can get their heads in cover make no to-do of the rest of them.

Of four-footed plagues the coyote is worst by numbers and incalculable cunning; and of him there is much that may be said to a friend able to dispense with the multiplication of instances.

In seventeen years a hill frequenter is not without occasion to listen at lairs when the sucking pups tumble about and nip and whine under a breath; to observe how they endure captivity among the wickiups or at some Greaser's hut; to fall in with them going across country and not be shunned, they understanding perfectly that skirts and a gun go infrequently together; to hear by night the yelping two-toned howl by which they deceive as to numbers, the modulations by which they contrive to make it appear to come from near or far, but never absolutely at the point from which it issues. And one has not to hear it often to distinguish the choppy bark by which the dog of the wilderness defies the camp from the long, whining howl that calls up a shape like his shape from the waste of warm, scented dusk.

On the high mesas when the thick cloud-mist closes on three sides of the trail, a coyote coming out of it unexpectedly trots aside with dropped head or turns inquiringly with a clipped noise in his throat like a man accosting a woman on the street before he is quite sure what sort she is, and may wish his hail to seem merely an inadvertence. But with all this, there is not the hint of any sound by which they talk comfortably together. Nothing passes between them but the fanged snarl when they fight, and the long, demoniac cry of the range.

Once when there was a pestilence among the rabbits so that they died in inconceivable numbers, lying out a long time on the bank of a wash under the Bigelovia to discern, if I might, the behavior of scathless rabbits toward those that were afflicted; lying very still toward the end of the afternoon, a coyote came down the wash, trotting leisurely with picked steps, as if he had just come from his lair, and not quite certain what he should be about. At that moment another crossed his trail at right angles,

trotting steadily as one sure of his errand.
They came within some feet of each other, the
nostrils of both twitched, they turned toward
each other with a look, long and considering
— ah, such a look as I had from you just now,
when I said that about the likeness of a man
to a coyote, intelligence deepening in the eye
to a divination of more than the fact says. And
at this look which hung in suspense for the
smallest wink of time, the one coyote fell in
behind the other and continued out of sight,
trotting with the same manner of intention
toward the same unguessed objective. Their
jaws were shut, no sound loud enough to be
heard at twenty feet passed between them ; but
this was open to understanding, that whereas
one of them before that look exhibited no sense
of intention, they were now both of the same
mind. And if we cast out all but the most
obvious, and say it signifies no more than that
one followed the other on the mere chance of
its being worth while, we are only the more at
a loss to account for all that they do to the
sheep.

Knowing the trick of frightened sheep to run

down hill and scatter as they descend, coyotes always attack on the lower side, and shepherds in a hill country camp below the flock to prevent them. Though seven is the largest pack I can attest to, they are reported to harry the sheep in greater numbers, and so rapid is the flash of intelligence between them that on the scattering of the flock, when one lamb or several are to be cut out, it is always by concerted action; and in longer runnings the relays are seen to be so well arranged for that no herder who has lost by them instances a failure that can be laid to the want of foreplanning. It is hardly the question whether coyotes in a raid will get any of your lambs, but how few.

Once slaughter is begun it is continued with great wastefulness unless arrested by the dogs. The coyotes understand very well how to estimate the strength of this defense, and finding attack not feasible, love to stand off in the thick dark and vituperate. No dog can forbear to answer their abuse with like revilings, and it is understood by them that when coyotes bark they do not mean thieving. Now this is most interesting, that the coyotes know that

they have made the dogs so believe. Not only
have they learned the ways of sheep and sheep
dogs, but also — and this is going a step beyond
some people — they are able to realize and play
upon the dog's notion of themselves. So on
a night when there is no sound from the flock
but the roll of the dreaming bells, warm glooms
in the hollows and a wind on the hill, three or
four of the howlers slip to the least assailable
side of the flock and there draw the dogs by
feints of attack and derisive yelpings. Then
the rest of the pack cut noiselessly into the
flock on its unguarded quarter and make a sufficient killing. And all this time the coyotes
have not said a word to one another.

A trick the herder has imposed on the sheep
by way of frustrating attack is to form the
flock with the heads all turned in, the dogs
being trained, on the hint of coyotes hunting,
to run about the closed herd and nip the fro-
ward members until the throats, the vulnerable
point, are turned away from the enemy. A
coyote will always be at considerable pains to
provoke a suitable posture for attack.

But there are no such killings now as in the

time when Jewett destroyed eight hundred coyotes in two years at Rio Bravo, and in all that time was unable to keep any dogs, so plentifully was the range spread with poisoned meat for the lean-flanked rogues.

They are still worst at the spring season when the young are in the lair and about the skirts of the mountains below the pines, for the snow prevents their inhabiting high regions except briefly in mid-season ; and on the plains where water-holes are far between they will not follow after the flocks, for meat-eaters must drink directly they have eaten.

Wanton killers as the coyotes are, one bob-cat can often work greater destruction in a

single night, for it comes softly on the flock, does not scatter it, kills quickly without alarm, and since cats take little besides the blood and soft parts of the throat, one requires a good bunch of lambs for a meal. Both cats and cougars have a superior cunning to creep into the flock unbeknown to the dogs, and the cougars, at least, go in companies; so if they manage not to stir the sheep and set the bells ringing to alarm the herder they get away unhurt with their kill. A cougar will hang about a flock for days, taking night after night a fresh wether of a hundred pounds weight, throwing it across his shoulder and carrying it miles to his young in the lair, with hardly so much as a dragging foot to mark his trail. It is chiefly by tracking them home or by poisoning the kill which the beast returns to, that the herder is avenged; for in the night lit faintly by cold stars, when the flock mills stupidly in its tracks with the cougar killing quietly in its midst, a gun is no sort of a weapon to deal with such trouble. Jewett reports four of these lion-coated pirates visiting his corral in a single night, each jumping the four-board fence and making off with a well-grown mutton ; and on another occasion the loss of sixty grown sheep in a night to the same enemy.

It is the conviction of most herders that all the slinking cat-kind are cowardly beasts, though stubborn to leave the kill unsatisfied, valuing their skins greatly, and even when attacked, fighting only to open a line of retreat. You will hear no end of incidents to convince you of this, but find if you swing the talk to bears that the herder's knowledge of them is like the ordinary man's understanding of wool tariff reforms, contradictious generalities in which he dares particularize only from personal experience. A bear, it seems, can, if he wishes, get his half-ton of weight over the ground with the inconceivable lightness of the wind on the herd grass; but he does not often so wish. He may carry his kill to his den or elect to eat it in the herder's sight, growling thunderously. He may be scared from his purpose by the mere twirling of your staff with shouts and laughter, and when he has gone a little way decide to return with wickedness glowing phosphorescently in the bottoms of his little pig's eyes, and grievously afflict the insulter. At one time the snapping of a wee bit collie at his heels sends him shuffling embarrassedly along the trail, and at another he sits back on his haunches inviting attack, ripping

open dogs with great bats of his paws, or snatch-
ing them to his bosom with engulfing and dis-
astrous hugs. He is not crafty in his killings,
but if he finds the mutton tender will return
to it with more bears, making two and three
flock-journeys in a night.

Singular, even terrifying, as evincing the
insuperable isolation of man, is the unaware-
ness of the wild kindred toward the shepherd's
interests, his claims, his relation to the flock.
The coyote alone exhibits a hint of reprisal in
that he neglects not to defile the corners of
the herder's camp and scratch dirt upon his
belongings, but to the rest he is, it appears, no
more than a customary incident of the flock, as
it might be blue flies buzzing about the kill.
All their strategies are directed toward not

arousing the dogs, man being uneatable, though annoying, not necessary to be closed with ex-cept in the last resort.

All these years afford me no more than two incidents of herders being damaged by beasts, one in Kern River having come to close quarters with a wounded bear which the dogs finally drew off, but not until the man's hurts were past curing. Yet in that region bears are so plentiful that they come strolling harmlessly across the recumbent shepherds in the night, or burn themselves with savory hot frying-pans lifted from the fire when the herder's back is turned. Or so it was in the days before the summer camper found that country.

At San Emigdio a she bear brought down her cubs on a moonless night to teach them killing, and Chabot, the herder, waked by the sound of running, hearing her snuffling about the flock, set on the dogs and himself attacked with his staff. This he would never have done had he been aware of the cubs, for though a grown bear suffers cudgeling with tolerable good humor, she will not endure that it should threaten her young. Therefore, Chabot carried the marks of that indiscretion to his grave.

But if you could conceive of the ravagers of the sheep-pens being communicative, it is plain that they would remark only, with some wonderment, but no recognition of its relativity, the irritating frequency with which man-things are to be found in the vicinity of flocks.

XI

THE SHEEP AND THE
RESERVES

CHAPTER XI

THE SHEEP AND THE RESERVES

WHEN the Yosemite National Park was first set apart, I said to a shepherd who was used to make his summer grazing there, —

"What shall you do now, Jacques?" — Jacques not being his real name, as you will readily understand, seeing the thing I have to relate of him. Jacques threw up his head from his hairy throat with a laugh.

"I shall feed my sheep," he said, "I shall feed them in the meadow under the dome, in the pleasant meadows where my camp is, where I have fed them fifteen years."

"But the Park, Jacques, do you not know that it is closed to the sheep and the whole line of it patroled by soldiers?"

"Nevertheless," said the shepherd, "I shall go in."

Afterwards I learned that he had done so, and at other times other shepherds had fed there, and at times the newspapers had a note to the effect that sheep had been caught in the Park Reserve and driven out. Sierra lovers who frequented the valley of falling waters came often upon fresh signs of flocks and spoke freely of these things, which, however, did not reach to places of authority. There was a waif word going about sheep camps, and now and then a herder who, when he had two thirds of a bottle of claret in him, was willing to make strange admissions.

"Five gallons of whiskey," said Jacques, "I pay to get in and take my own chance of being found and forced out. We take off the bells and are careful of the fires. Last year I was in and the year before, but this summer some fools going about with a camera found me and I was made to travel. Etarre was in, and the Chatellard brothers."

" And did these all pay ? "

" How should I know ? They would not pay unless they had to. But it is small enough for two months' feed; and if the officers found us we had only to move on."

All the gossip of the range is by way of proving that the shepherd spoke the truth. It is not impossible that the soldiers despised too much the work of warding sheep off the grass in order that silly tourists might wonder at the meadows full of bloom. The men rode smartly two and two along the Park boundary; one day they rode forward on their appointed beat and the next day they rode back. Always there was a good stretch of unguarded ground behind them and before. If they found tracks of a flock crossing their track they had no orders to leave the patrol to go after it; they might report — but if it were made more comfortable not to ? This is not to say that all the enlisted men of that detachment could be bought, — and for whiskey too! But in fact a flock can

cross a given line in a very narrow file, and it
was not necessary that more than two or three
of the patrol should be complaisant.

During the Cuban war, the military being
drawn off for a business better suited to their
degree, and the Park left to insufficient war-
dens, the sheep surged into it from all quarters.
They snatched what they could, and when
routed went a flock-length out of sight and
returned to the forbidden pastures by a secret
way. I dwell upon this, for it was here and by
this mismanagement that the foundation was
laid for the depredations, the annoyance, and
misunderstanding that still make heavy the
days of the forest ranger.

After the return of the soldiery, enforce-
ments were stricter but trespasses made more
persistent by a season of dry years that shortened the feed on the outside range. The sheepmen were not alone in esteeming the segrega-
tion of the Park for the use of a few beauty-
loving folk, as against its natural use as pasture,
rather a silly performance. No proper penal-
ties were provided for being caught grazing on
the reserve. An ordinance slackly enforced is

lightly respected. More than that, sheepmen who had by long custom established a sort of right to those particular pastures considered themselves personally misused. They must now resort to infringement on the grazing rights of others or be put out of business; not, however, before they had made an effort and a tolerably successful one, to break back to the forbidden ground.

All this time there were going on in California remote and incalculable activities that should turn the general attention at last toward the source of waters. One feels perhaps that we affect to despise business too much; it is in fact the tool by which the commonalty carves toward achievements too big for their understanding, which they laugh at while forwarding. At this time and for some years before, in all the towns of the San Gabriel and the San Joaquin and the coastward valleys there were men going about on errands of the business sense, seeing no farther than their noses, perceiving no end to their adventure other than the pit of their own pockets, denying and not infrequently contriving against the larger purpose which they served. The bland Promoter who sold irrigable lands for a price that made the buyer gasp, and while he was gone around the block to catch his breath raised it a hundred per cent, hastened, though unaware, the conservation of the natural forests. Incidentally he worked the doom of the hobo herds.

It is fortunate that the heads of government, like the tops of waves, move forward under pressure of an idea at rates much in advance of the common opinion. The breaking of that surge toward forest preservation was in a line about the chief of the watersheds beyond which it was not lawful for sheep or cattle to pass. Here in my country it cuts off squarely south of Havilah, runs straightly north to the spur of Coso Hills, where the desert marches with it past Olancha, trends with the Sierras north by west past Lone Pine, past Tinnemaha, past Round Valley and Little Round Valley, and turns directly west to meet the Yosemite Park. Returning on the other slope, it encompasses the Northfork country, the country of Kaweah, the sugar-pine country, and the place of the sequoias, Tule River, Kings and Kern, all the noble peaks that rear about Mt. Whitney and the pleasant slopes of Three Rivers and Four Creeks, in short all that country of which I write to you.

I said that at first neither sheep nor cattle might pass it, but very shortly it was granted that cowmen living near the reserve should, by special permit, feed their stock on certain of the most generous meadows at the set time of the year. It is not to be wondered at that the sheepmen conceived this a blow directed at the wool and mutton industry, and finding the price of stock sheep forced down by these measures, excused their trespasses by their necessities. Some there were who slipped in by night and slipped out, ashamed and saying nothing, others who infringed boldly and came out boasting, as elated, as self-gratulatory as if they had merged railroads or performed any of those larger thieveries that constitute a Captain of Industry.

There was a Basque, feeding up and down the Long Trail, who was notably among the offenders. A trick of his which served on more than one occasion was to start a small band moving, for he had fifteen thousand head, and having attracted the ranger's attention by boasts and threats made with the appearance of secrecy, in places most likely to reach the ranger's ear, to draw him on to following the decoy by suspicious behavior. Then the Basco would bring up the remainder of the flocks and whip into the Reserve behind the ranger's back. Once a day's journey deep in the Sierra fastnessess, it would be nearly impossible to come up with him until, perhaps, he neared the line on his fall returning. The sheepmen had always the advantage in superior knowledge of the country, of meadows defended by secret trails and false monuments, of feeding grounds inaccessible to mounted men, remote and undiscovered by any but the sheep. They risked much to achieve a summer's feeding in these fair, inviolate pastures. The most the rangers could do against them was to scatter and harry the flock so as to make the gathering up difficult and expensive. The business was also hindered by the inadequacy of the ranger force. Every man had more territory than he could well ride over, and rode it fast at the

end of a red tape centred in Washington, D. C.
The service did not know very well what it
wanted, and the pay was much below the price
of the fittest men. Whatever the ranger did
was at the mercy of the man at the other end
of his tape, who like enough had never seen
a forest off the map. Whatever went on, the
ranger reported in a detailed account of each
day's proceedings. After which he explained
the report. If the Tape Spinner wrote back to
know why on a given day he had but covered
the distance between two places no more than
five miles apart on the map, and the next day
had ridden fifteen, no matter what was doing

in the way of trespass or forest fires, the ranger paused politely to explain that the first day's riding was pretty nearly straight up in the air, over broken ground, and the second through a pleasant valley. Still, if the explanation failed to satisfy, the forester's pay was docked.

On one occasion a ranger saw against the morning sky the pale saltire of forbidden fires at a time of the year when forest fires were most to be abhorred. Two days' hard riding discovered the fire to be in a small granite fenced basin, nearly burnt out with its own fury. He so reported and had his pay cut for the whole time of his fruitless errand. But suppose the fire had not been in an isolated basin, and suppose he had not gone to see? Another ranger requiring powder for blasting a landslip from a ruined and impassable trail, went to the nearest town, which happened to be a day's ride from the reserve. Timidly he submitted the bill for the powder and it was allowed, but the man was cut two days' pay for being out of the Reserve without leave. I could tell you more of these absurdities, but I am ashamed of them; besides, the sense of the ser-

vice is always toward greater efficiency; moreover the sane, inspiring work of forest preservation sweeps to its larger purpose not too
much hindered by the fret of departmental
inadequacies. But when these things are so,
you can understand that the herders could the
more easily take the advantage.

I shall not here recount the whole of that
struggle between the rangers and the sheep,
the experimental kindnesses, the vexed reprisals, the failures, triumphs, and foolish heroisms.
It is true that not all the keepers of sheep
forged over the viewless line of the Reserve
unless it might be by inadvertence, for in the
beginning it was not very clearly determined.
Respectable sheep-owners sat at home and
ordered their herders to bring fat mutton and
full fleeces back from the curtailed pastures.
These simple-hearted little men came near to
achieving the impossible. Those who would
have done nothing on their own behalf stole
stoutly in the interest of their owners. One
caught at it would have shot the ranger, only
the ranger shot first. And if their very dogs
were not in league with them, how is it that

the flock of Filon Gerard stampeded so for-
tunately as they were crossing, under escort
of the rangers, at Walker's Pass. True, Filon
had been kept hanging about the Pass on the
barren mesa for several days, waiting for the
arrival of the escort, and the narrow strip of
crossing allowed was already eaten off to the
grass-roots by earlier passing. No doubt the
sheep then were crazed by hunger, as Filon
avowed. It seems certain that some signs
passed between him and the dogs at the moment of stampeding; and by the time the
ranger had helped to gather them up they had
all a fill of the fresh, sweet grass.

When Jean Rieske camped where he had
been wont to rest on his passage up from Mojave, over-tired, with a footsore, hungry flock, —
for he had attempted the passage too early, before the desert feed was well advanced, — when
he had no more than lighted his fire to warm
his broth, it being then long past dark, down
came the rangers upon him with orders to move.
For what? A new regulation; that was all
they knew. Three days ago it had been lawful
to camp in this place, now it was not. Jean

Rieske moved on. There were some miles to cover to another camp, the season was early, and the lambs were young; in the darkness, fatigue, and confusion they became separated from the ewes. The rangers were also tired, cold, and hungry, and harried unnecessarily the flock. Nights on these high mesas the keen still cold bites to the bone — and Jean Rieske could not carry all the lambs of one flock in his bosom. What indeed are half a hundred lambs to the letter of the law?

There was a ranger rode out of town to pass over the gap between two bulky, grey, and wintry mountain heads, in the month of frequent rains ; and a mile over the line of the Reserve came upon a Portuguese herder of two thousand blackfaces, working straight toward the lake basins of the ten thousand foot level. He turned the man back, saw the sheep out of bounds, watched them dip away, the herder still protesting the virtue of his intention, into a hollow where there was thick black sage, and urged by his errand, pricked forward on the trail. Even with this delay he hoped to make the pass and the meadow of Bright Wa-

ter by night, but when he had come to the first
of the lingering drifts he found the trail choked
with rubble, and just beyond, obliterated in a
long, raw scar where the whole front of the
hill, made sodden by recent rains, had sloughed
away into the cañon below. This sent him
back on his tracks in time to find the same
herder working industriously over the same
ground from which he had been routed earlier
in the day. The ranger told me afterward
with great relish how he pulled his gun — in
this country when we say gun we mean a six-
shooter — and drove the Portuguese down the
trail before him. I am told there are places on
that grade where a man in a hurry may cover
as much as twenty feet without hitting the
ground. The flock was all of that year's increase, lately weaned and not yet flock-wise;
they began to drop behind on the steep, in the
pitfalls of the strewn boulders, in the stiff wat-
tles of the chaparral. The ranger and his man
came out of the Reserve at a flying jump, where
the ranger breathed his horse and the Portuguese lay on the ground, bellowed with anger,
and tore up handfuls of the scant grass.

In the midst of rage and trickery there were two who knew nothing of it, but remembered only their devotion to the flock. At the last it was in pity for the incredible great labors of the dogs who covered, with tongues out and heaving sides, the broken steeps of the cañon so many times in the breathless afternoon, that the ranger permitted the herder to get upon his feet and gather the remnant of the flock. I should say that the fellow lost the half of the year's increase by that venture. And no longer ago than the time when every swale of the long mesa overflowed with the blue of lupines, as blue as sea water, the rangers found a shepherd feeding on the tabooed ground. He said, and the rangers believed him, that he was not aware of trespass. Nevertheless, as their orders ran, they began to drive the sheep outward, scattering as they went. The little Frenchman wearied himself to keep them close, he was fit to burst with running, he sobbed with the laboring of his sides, tears streamed from him ; and when at last he was able to send hired men to gather up his flock, it had cost him as much as a whole summer's feed in fenced pastures.

" And all the time," said the ranger, " I was
perfectly sure that he had crossed the line
without knowing it, as he might easily have
done, for there were no monuments at that
place." I confess to a great liking for these
lean, keen, hard-riding fellows, who have often
an honest distaste for the orders they execute
with so much directness and simplicity, and
from whose account it appears that the law at
times out-does itself, and, thinking to prevent
infringement, inflicts a damage.

Do not suppose I shall enter a proof or a
denial of all the sheep have done to the water-
sheds, what slopes denuded, what thousand
years of pines blackened out with willful fires.
These things have been much advertised with
all the heartiness and particularity of those sure
of the conclusion before the argument is in-
itiated. I might add something to the account,
instancing the total want of young shrubs of
the bitterbrush, the *wheno-nabe* of the Paiutes,
purshia tridentata of the botanist, greedily
sought by sheep and cattle. This extraordinarily bitter-savored shrub of dark green, shining, small foliage, has a persistent bark, brown

and fibrous, grown anew every year, half sloughed away so that a stem might display an inch or more of this shaggy covering, strong as hemp, which the Indians of old time shredded and wove into mats for lining their caches and storing pine nuts against need. No vermin attacked it, nor rot nor dampness. Two of these mats I have, taken from a cache in the Coso hills, forgotten as long ago as before the white man inhabited there, which was before the Gunsite mine was lost or ever Peg-Leg Smith had made his unfortunate " passear ; " and the fibre is yet incredibly fresh and strong. But when the Indians discovered cloth and canned goods so much more to their taste, then the demand for the *wheno-nabe* fell off and the strip of country where it grew became part of the Long Trail. Normally the plant should have increased in those years, but when after an interval it was thought possible to reinstate the ancient craft, the sheep and cattle had left us no plants of the bitterbrush in that neighborhood but such as appeared as old as the Indians who remembered the knack of its use.

Also I could say something of the hills be-

hind Delano that once were billowy and smooth
as the backs of the ocean swell, and after so
many years of close-herded sheep trampling in
to the annual shearing are beaten to an impervious surface that sheds the rain to run in hollows and seam them with great raw gullies so
that the land shows when the pitiless high light
of noon searches it, like the face of an old
courtesan furrowed with the advertisement of
a too public use.

You will find the proof of things like that in
the government reports, together with many
excellent photographs of before and after, to
convince you of the plague of sheep. For you
notice, curiously, all this anathema is directed
against sheep, whereas we who have followed
after the bells know that it is to be laid to the
sheepman, and to a sort of sheepman fast disappearing from the open range. What I mean
to say, while admitting the damage, is that
there is nothing, practically nothing, in the nature of sheep inimical to the young forests or
the water cover. Is it not the custom otherwhere to put sheep on worn-out lands to renew

them? Have not flocks been turned to the vine-
yards to lighten the pruning? Does any farmer
complain who has hired his alfalfa fields to the
herders, or manure them other than with the
droppings of the sheep? Do sheep eat young
pines except of starvation, or crop the grasses
into the root-stock, or trample the earth into
a fine dust, or break down the creek banks in
passage except the herder imposes such a necessity? Do sheep light forest fires or turn
streams from their courses?

But suppose you have man laying his will
heavily on the flock, a man say who has a wife
or a sweetheart in France and looks in six or
seven years to sell out and go back to her,
knowing nothing of the ultimate disaster, car-
ing nothing for those who come after him.
Such an one with sheep under his hand can
use them to incalculable damage. It needed
some illuminating talks with a man who had
run his stock on the fenced pastures of Men-
docino to get this matter fairly into shape.
Shepherds who feed on their own ground blame
only themselves if their pastures deteriorate,
and they chiefly suffer for it. Seeing how all

creatures so use the face of the earth to better
it, it is ridiculous to suppose that sheep left
reasonably free from man-habits and not
encouraged to increase in excess of the feed
produced, should incontinently work us harm.
They clean up the dry grass and litter by
which the smouldering fire creeps from pine
to pine; ranging moderately on the hillslopes
they prune the chaparral which by smothering
growth and natural decay covers great areas
with heaps of rubbish through which the shrub
stems barely lift their leaf crowns to the light
and air. Frequently in such districts after a
fire, trees will spring up where no trees were
because of the suffocating growth.

There is always a point beyond which it is
not well to push any native industry to the
wall. Consider what the price of wool and
mutton must grow to be when these are raised
on irrigated lands. But what if it were granted
to sheepmen as to cattlemen for a small rental
to graze on the withdrawn pastures under proper
circumstances of supervision? As to this matter there is much that wants learning. What
the forester must know is the precise time be-

tween the two nodes of the year when grazing
is accomplished without harm to the water
cover. As to the first, when the annual grasses
begin to stool in the spring, before their roots
are established, when they perish from a single
cropping; as to the last, the hour beyond which,
if cut off in mid-stem, they ripen no seeds. He
is to choose also the times of moving from
meadows across the forested lands. Fortun-
ately the wild pastures are still deep under
stained, sludgy snows when there is over all
the leaves of the pine, the burnished bloom,
the evidence of the rising sap, at what time
a break or a scar retards the season's growth.
But a little later than the time when rains begin, the forces of life and death are so evenly
balanced that the rake of the sharp hoofs
downward, still more the impact of the heavy
tread of the steers, jars out the little dryad of
the sapling tree. It sticks in my mind that there
is not enough attention paid to the moving of
cattle through the pine woods in the climac-
teric of the year.

It is an instance of how the right conduct
of any business forces itself on those who con-

cern themselves about it with an open mind,
that no longer ago than the time when this
book began to shape in my mind, there was no
forester but regarded the sheep with abomination, and now none, in my district at least,
otherwise than generously inclined toward the
properly conducted flock. Though it is not
often and so completely that one is justified in
the comfortable attitude of having known it
all the time.

XII

RANCHOS TEJON — SOME AC-
COUNT OF AN OLD CALIFORNIA
SHEEP RANCH AND OF DON JOSÉ
JESÚS AND THE LONG DRIVE.

CHAPTER XII

THIS year at Button Willow they sheared the flocks by machinery, which is to say that the most likable features of the old California sheep ranches are departing. That is why I am at the pains of setting down here a little of what went on at the Ranchos Tejon before the clang of machinery overlays its leisurely picturesqueness.

When Mexico held the state among her dependencies she gave away the core of it to the most importunate askers. A good lump of the heart land went in the grants of La Liebre, Castac, and Los Alamos y Agua Caliente, to which Edward Fitzgerald Beale added in '62 the territory of the badger, called El Tejon. This principality is three hundred thousand acres of noble rolling land, lifting to mountain summits and falling off toward the San Joaquin where that valley heads up in the meeting of the Sierra and Coast Ranges. The several grants known as Ranchos Tejon dovetail together in the high, wooded region where the Sierra Nevadas break down in the long, shallow passage of Cañada de las Uvas.

Beginning as far south as the old Los Angeles stage-road, which enters the grant at Cow Springs, the boundary of it passes thence to Tehachapi ; northward the leopard-colored flank of Antelope Valley heaves up to meet it. Here begins the Tejon proper, crossing the railroad a little beyond Caliente, encompassing Pampa on the northwest; from hence trending south, stalked by blue mirages of the San Joaquin, it divides a fruitful strip called since Indian occupancy the Weed Patch, and coasts the leisurely sweep of the Sierras toward Pastoria. This guttering rift lets through the desert winds that at the beginning of Rains fill the cove with roaring yellow murk. About the line of the fence, bones of the flock over-blown in the wind of '74 still stick out of the sand. Hereabout are the cleared patches of the homesteaders, where below the summer limit of waters the settlers play out with the cattlemen and the sheep the yearly game of Who Gets the Feed. Thence the boundary runs west to Tecuya; here the oaks leave off and the round-bellied hills of San Emigdio turn brownly to the sun. Castac, which is to say The Place of Seeping Springs, basks obscurely in the shallow intricacies of cañon behind Fort Tejon, finding the border of La Liebre a little beyond the brackish lake, wholly to include the ranch of the cottonwoods and warm water, otherwise Los Alamos y Agua Caliente. Beginning at Pampa, a fence rider should compass the whole estate in a week and a day.

For those so dry-as-dust as to require it there is an immense amount of stamped paper to certify the time and manner of Beale's

purchases, but I concern myself chiefly with
the moment when he married the land in his
heart, coming first out of the dark, tortuous
cañon of Tejon, not the fort cañon, but that
one which opens toward the ranch house, and
looked first on the slope and swale of the bask-
ing valley. If it is yet called the loveliest land,
judge how it looked to him after the thirsts,
the vexations, the epic fatigues of his explora-
tion of the thirty-fifth parallel. Back of that
lay San Pascual, the figure of himself as a
swarthy young lieutenant carrying to Wall
Street the news and the proof of the first
discovery of gold; and through a coil of high
undertaking as a bearer of dispatches looping
back to the day when President Jackson saw
him fight out some boyish squabble in the
streets of the Capital and appointed him to
the Navy.

" The boy is a born fighter," said Old Hick-
ory, " let him fight for his country." He was
not the less pleased when he learned that the
lad was a grandson of Commodore Truxton
whom the President had admired to the extent
of naming a race-horse after him.

It was all a piece of the simplicity of the time that grandmother Truxton, when she heard of the appointment, cut the buttons off the dead Commodore's coat to sew on the midshipman's jacket, so that the boy arrived at the frigate Independence wearing that insignia, whereat the other middies laughed. Something less than a score of years stretched between the time when the boy of twelve lay miserably in his berth contriving how to get rid of the Commodore's buttons and the time when he rode with Frémont into the full-blossomed Tejon; but if you said no more of them than that they had sharpened and shaped the man for knowing exactly what he wanted and being able to get it, you would have implied a considerable range of experience.

Knowing about San Pascual, you conceive that the man must have had extraordinarily the faculty of dealing with primitive peoples. I suppose that Beale was the first official to discover, or to give evidence of it, that it is wiser for Indians to become the best sort of Indians rather than poor imitation whites. That part of the estate known as Rancho el

Tejon had been an Indian Reservation, gathering in broken tribes from Inyo, from Kern and Tule rivers and Whiskey Flat, prospering indifferently as Indians do in the neighborhood of an idle garrison such as Fort Tejon. Beale, being made Superintendent of Indian Affairs, began to prove the land and draw to him in devotion its swarthy people, and the Reservation being finally removed to Tule River, there passed to him with the purchase of El Tejon, the wardship of some dozens of Indian families. Such of them as longed homesickly for their own lands melted from Tejon like quail in nesting time, by unguessed trails, to the places from which they had been drawn, and to those remaining were accorded certain rights of home-building, of commons and wage-working, rights never abated nor forsworn during the lifetime of Edward Beale.

There were notable figures of men among these Tejon Indians; one Sebastian whom I have seen. Born a Serrano in the valley of San Gabriel, he was carried captive by the Mojaves, one spark of a man child saved alive when the hearth fires were stamped out in war. He being an infant, his mother hid him in her bosom; with her long hair she covered him; between her breasts and her knees she suckled him in quietness until the lust of killing was past. Among the captive women he grew up, and escaping came to know the country about Kern River as his home. Here when Frémont came by, exploring, the river was at flood, a terrible, swift, tawny, frothing river, and no ford. However, there was Sebastian. This son of a chief's son stripped himself, bound his clothing on his head, swam the river, brought friendly Indians, made fast a rope across, brought the tule boats called "balsas," ferried over the explorers, and got from Frémont for his pains — nothing; a rankling slight until the old man died. But between Sebastian and Beale grew up such esteem from man to man as lasted their lives out in benefits and devotion.

One finds tales like this at every point of contact with the Tejon, raying out fanwise like thin, white runways of rabbits from any water-hole in a rainless land. The present master of the estate has told me, himself all unaware, and

I secretly delighted to see the land rise up and grip him through the velvet suavity of years, how when he was a boy and the court between the low adobes closed at night as a stockade, red eyes of the Indian campfires winked open around the swale where the ranch house sat, and at the end of the first day's drive toward Los Angeles, as they would ride at twilight over the Tejon grade, the circling fires blossomed out from the soft gloom, watching on their trail. More he told of how he went up the cañon, full of little dark bays of shadow, with his father to bury old Nations, of how the dead mountaineer looked to him through the chinks of the cabin, large in death, and how being no nearer than sixty miles to a Bible, the General — he was Surveyor-General at one time — contrived a ceremony of what he could remember of the burial service, and the Navy Chaplain's prayers, and the tall, hard-riding Texans and Tennesseeans, clanking in their spurs, came down to be pall-bearers, lean as wolves drawn from hollows of the mountains as lonely as their lairs.

I should have said that, inside of the ranch

boundaries, there were sections and corners of government land, these drawing to them, by election, westward-roving clans of southern mountaineers. Here they brought the habits of freedom, their feuds, yes, and the seeds of the potentialities that make leaders of men. Here grew up Eleanor and Virginia Calhoun, nourished in dramatic possibilities on the drama of life. I remember well how Virginia, during the rehearsals of Ramona, when we milled over between us the possibilities of what an Indian would or would not do, broke off suddenly to say how clearly the peaks of Tejon would swim above the middle haze of noon, or how she had waked mornings to find the deer had ravaged the garden, or a bear in her playhouse under the oaks.

But the real repository of the traditions of Tejon is Jimmy Rosemeyre, — and in the West when a whole community unites to call a man by his first name, it is because they love and respect him very much. Jimmy, who crossed the plains in '54, and was drawn down from Sacramento by natural selection to Tejon; Jimmy, who, because of his comeliness among so many dusky folk, was called Jimmy " *werito*," Jimmy the Ruddy; who, when he had a good horse under him, a saddle of carved leather-work, *botas*, deep-roweled spurs and a silver-trimmed sombrero, knew himself a handsome figure of a man; James Vineyard Rosemeyre, who saveys the tempers and dispositions of men, who knows the Tejon better than its own master, the man whose hand should have been at the writing of this book.

It is well here to set forth the shape of the land, to know how it colors the life that is lived in it. Between the point of San Emigdio and the Weed Patch there is a moon-shaped cove, out of which opens, westerly, the root of the cañon by which Frémont and Kit Carson came through. The ranch house sits by the water that comes down guardedly between tents and tents of wild vines. Below the house by the stream-side the Indian washerwomen paddle leisurely at the clothes and spread them bleach-ing in the sun. Silvering olives and mists of bare fig branches slope down to the blossomy swale; deep in the court between the long adobes, summer abides, and yearly about the fence of the garden the pomegranates flame. The beginning of all these, and the oranges, Jimmy Rosemeyre brought up from the Mission San Fernando, going down with two live deer in a wagon and returning with cuttings and rooted trees. Six miles up the cañon are the adobe huts and the *ramadas*, the bits of fenced garden that make the Indian  rancheria. Rising out of laps and bays of the oak-furred ridges, pale smoke betrays the hearths of the mountaineers.

Below the ranch house in a wet spring the land flings up miles of white gilias and forget-me-nots, such as the Spanish children call *nievitas*, little snow; spreads on the flowing hill bosses the field of the cloth of the dormidera, collects in the hollows pools of purple wild hyacinth, deep enough to lie down in and feel the young wind walk above you on the blossom tops. Days of opening spring the cove is so full of luminosity that the backs of crows flying over take on a silver sheen. You sit in the patio when the banksia rose sprays out like a fountain, and hear the olives drip in the orchard; awhile you hear the stream sing and then ripe droppings from the young full-fruited trees. At night the hills are silent and aware, and all the dreams are singing.

Straight out from the ranch house runs the road to Castac and La Liebre. It turns in past the house of José Jesús Lopez, and runs toward Las Chimeneas. Here, to the left, is the camel camp. Nobody much but Jimmy Rosemeyre and the Bureau of Animal Industry knows about the camels that the government, by the hand of Lieutenant Beale, undertook to domesticate on the desert border. Twenty-nine of them, with two Greeks and a Turk, came up by way of The Needles, across the corner of Mojave to Tejon. There I could never learn that they accomplished more than frightening the horses and furnishing the entertainment of races. They throve, — but no American

can really love a camel. Whether they admit it or not, the Bureau of Animal Industry is balked by these things. Nothing remained of them at Tejon but tradition and a bell with the Arabic inscription nearly worn out of it by usage, cracked and thin, which Jimmy Rosemeyre, in a burst of generosity, which I hope he has never regretted, gave to me. Hanging above my desk, swinging, it sets in motion all the echoes of Romance.

The road runs whitely by Rose's Station. Los Angeles stages used to stop there, but I like best to remember it as the place where Jimmy Rosemeyre had a circus once, in the time when circuses traveled overland by the stage-roads from camp to roaring camp. Never was a more unpromising quarter than this

tawny hollow with one great house bulking darkly through the haze. But Jimmy wanted to see that circus.

"You go ahead with the show," said he, "I'll get the crowd;" and he sent out riders. No lean coyote went swiftlier to a killing than word of the circus went about the secret places of the hills. The crowd came in from Teha-chapi, from Tecuya and San Emigdio and the Indian rancherias; handsome vaqueros with a wife or a sweetheart before them in the saddle, — and that was the time of hoopskirts too, — Mexican families with a dozen or fifteen mu-chachos and muchachitas in lumbering ox carts, squaws riding astride with two papooses in front and three behind. They brought food and camped by the waterside, sat out the after-noon performance, and after feasting returned with unabated zest at night. But in the year I spent at Rose's Station I found nothing better worth watching than the antelope that signaled in flashes of their white rumps how they fared as they ran heads up in the golden amethyst light of afternoon.

The road climbs up the grade from the foot

of which trends away the ineffaceable dark line
of the old military road, visible only from the
heights as the trail of forgotten armies from
the summits of history. It leads to the ruins
of Fort Tejon, built under the sprawly old oaks
where the cañon widens, costing a million dollars and accomplishing less for the pacification
of the Indians than one Padre, says Jimmy
Rosemeyre. Across the brook from the road,
across the meadow of yerba mansa, across the
old parade-ground, at the lower corner of the
quadrangle of ruined adobes is the Peter Lebec
tree. Under it the first white man died in that
country and under it the first white child was
born. General Beale himself showed me the
great bough that was lopped away to rid the
woman of fear of its overhanging weight when
she came to her distressful hour. Lebec, I
spell it now as it was rudely carved in the inscription, was buried in 1837, and after more
than fifty years, by the rediscovered inscription
printed in reverse on the bark grown over the
blaze, and by exhuming of the body was proved
the current Indian tradition that while he lay
under it, heavy with wine, and his camp-mate

away hunting, a bear came down out of the oak and partly devoured him.

You get more than enough tales of killings and wickedness hereabout, bandit tales of Mason and Henry, and Vasquez the hard rider. I could show you the place by the dripping spring where I found the pierced skull, — pleasanter to walk in the white starred meadow and hear tremulous, soft thunder of wild pigeons in the oaks, to wind with the road's windings up the summit to Gorman and see the shadows well out of the cañons and overflow the land and the lit planets flaring low above the glade that holds the ranch house of La Liebre. This was the end of the second day's driving, when one went from Tejon to San Francisco by way of Los Angeles and the sea. The present lord of the Ranchos Tejon would follow this road with reminiscences past Elizabeth Lake, through San Francisquito cañon, clothed on with stiff chaparral, lit by tall candelabra of the Spanish bayonet, as far as the stark old Mission San Fernando with Don Andreas Pico bowing open the door and an Indian servitor in a single garment behind each

chair of the hospitable board. But he could go
as far as that without getting away from the
spirit of Tejon which
in General Beale's life
much resembled the
best of mission times.
The measure of regard
which he won from
the Indians was paid
for in respect for usages
of their own; as you
shall hear and judge in
the case of the *Chisera*.

A *Chisera* you must
know is a witch, in this
instance a rainmaker. In a dry year the General put the Indians to turning the creek into
an irrigating ditch to water the barley. Said
they: —

" Why so much bending of backs and breaking of shovel handles? There is a woman at
Whiskey Flat who will bring rain abundantly
for the price of a fat steer."

" Let her be proven," said the General, like
Elijah to the prophets of Baal.

The *Chisera* wanted more than a steer, — beads, calico, the material for a considerable feast, all of which was furnished her. First the Indians fed and then the *Chisera* danced. She leaped before the gods of Rain as David before the Ark of the Lord when it came up from Kirjath-jearim; she stamped and shuffled and swung to the roll of the hollow skins and rattles of rams' horns; three days she danced, and the Indians sat about her singing with their eyes upon the ground. Day and night they sustained her with the whisper and beat of their moaning voices. Is there in fact a vibration in nature which struck into rhythm precipitates rain, as a random chord on the organ brings a rush of tears? At any rate it rained, *and* it rained, and it *rained!* The barley quickened in the field, a thousand acres of mesa flung up suddenly a million sprouting things. Rain fell three weeks. The barley and the wheat lay over heavily, the cattle left off feeding, the budding mesa was too wet to bloom.

"For another steer," said the *Chisera*, " I will make it stop."

So the toll of food, and cloth, and beads was paid again, and in three days the sun broke gloriously on a succulent green world. It is a pity, I think, that the *Chisera* is dead.

Under the General's patriarchal hand there was never any real difficulty with the Indians at Tejon, though there was an occasion once at shearing-time, when there came out of Inyo a Medicine Man who gathered the remnant of the tribe to him at Whiskey Flat. He was credited with an unfailing meal-sack and promised healing to the sick, the maimed, and the blind. No doubt the easily springing hope of such as this augurs to the primitive mind its possibility. Whispers of it ran with the click of the shears in the sheds. Question grew into conviction and conviction to a frenzy. Useless to argue that these things, if true, would keep and the shearing would not; man after man, they dropped their shears with the unclipped merinos, and for this defection, a serious hindrance when no workers were to be had for sixty miles, they were never taken back into employment.

It was against this background of wild
beauty, mixed romance, and unaffected sav-
agery, that the business of wool-growing went
on at Tejon much as I have described it for
the Open Range, though running a flock on
patented lands lacks the chance of adventure
that pertains to the free pastures. It was
Jimmy Rosemeyre who brought the first
sheep to the territory of the badger, having
purchased as early as '57, a band of mustang sheep driven up from Mexico by Pablo
Vaca and Joaquin Peres, shaggy and unbidable little beasts that must be herded on horseback. Afterward he sold them to Beale, and
when by improvement of the breed they grew
tractable, the herding fell to the Indians.
Threescore herders in the best of times went
out with the parted flocks, and at that time
when the grass on the untrampled hills ripened
its seeds uncropped through successive years,
the feed grew shoulder high for the sheep.
The head shepherd moved them out from the
shearing like pieces on a board; mostly they
could make stationary camps, feeding out cir-
clewise for weeks at a time.

The sheep had no real enemies at Tejon but drouth and the bears. Against the drouth, the *Chisera* being dead, there was no remedy. The tale of the flocks was very strictly kept; every herder was required to show the skins of all that he killed or that were slain by beasts, or such as died of themselves, and in the driest year the number reached twenty-two thousand head. In '76, all the earth being sick with drouth prolonged, the fifty-eight thousand sheep were turned out in December unshepherded, the major-domo being at the end of contrivances for saving them alive. They sought the high places among the rocks, the secret places of the most high hills, and no man spied on their distresses. Being so trusted, the land dealt with them not unkindly, for when the first rains of October drove them to the foothills there were gathered up, of the original flock, fifty-three thousand. But in good years they saved all the increase, and made good with equal killings the ravages of beasts.

There were once great grizzlies at Tejon, but mostly the bears are of the variety called black by scientists because they are dark brown, or even reddish when the slant light shows them feeding on the mast under the oaks or gathering manzanita berries on the borders of hanging meadows, wintry afternoons. Black enough they look, though, lumbering up the trail in the night or bulking large as their shadows cross the herder's dying fire. Pete Miller is the official bear-killer of the Ranchos Tejon, though his account of the killings are as short as the items in a doomsday book.

"Tell me a bear story, Pete," say I, sitting idly in the patio about the time of budding vines. Says Pete, —

"Up here about three mile from the house there was a deef old Indian saw a bear going into a hollow tree; he heaved a chunk of fire in after him and shot him with a six-shooter when he came out."

The stamp of simple veracity is in Pete's open countenance.

"Another time," he said, "there was a bunch of bears up the cañon stampeded the sheep so they piled up in a gulch. No 'm, they won't anything but a gulch stop sheep once they get a-running; they was about two hundred of them killed. Me and two other fellows went up the next night — yes 'm, bears they always come back. We got the whole bunch. They was six." Pete sat on the edge of a chair and told tales like that for an hour. They all began with a bear getting after the sheep, and ended with Pete getting the bear.

"How many bears have you killed, Pete?" say I.

"I fergit, exactly," says Pete, fumbling embarrassedly with his hat; but current tradition makes it near to three hundred.

Nearly everybody at Tejon can tell a creditable bear story; this from Jimmy Rosemeyre, not to be behindhand.

"I went up to Plaza Blanco to see a herder," said he; "I was packing some venison on my horse; yes, you can put a deer on a horse if you blindfold him. The herder was toasting some strips of meat on a stick.

"'What's that?' said I.

"'Cougar,' he says, 'it's better than venison.'

"Thinks I, I'll try it, so I let my deer be and went to toasting pieces of cougar on the coals.

It was. Good and sweet. The herder was sleeping in a tapéstre — that's a bed on a platform in a tree. He said the bears bothered him some. But he was an all-right fellow; he wanted me to sleep in the tapéstre and let him sleep on the ground. Along in the night we heard the sheep running. It was dark as dark, a thick dust in the corral, and big lumps of blackness chasing around among the sheep. We could n't see to shoot, but there were oak poles smouldering in the fire. We whacked the big lumps over the head with them. Leastways we aimed to whack 'em on the head, but it was pretty dark. I guess we scorched 'em considerable by the smell. There was one wallowed in the creek to put himself out. Seemed as if that corral was full of bears, but in the morning when we counted the tracks there were only four."

But think of knowing a man who could whack four big California bears over the head with a fire-brand!

There was never anything to equal the spring shearing at old Tejon; when there were eighty thousand head to be clipped, you can imagine it was a considerable affair. Seventy-five or

eighty Indians bent backs under the sheds
for five or six weeks at a time, and Nadeau's
great eight-ox teams creaked southward to Los
Angeles, a hundred and twenty miles, with the
wool. All this finished with a fiesta lasting a
week, with prizes for races and cockfights, with
monte and dancing, and, of course, always a
priest at hand to take his dole of the shearing
wage and confess his people where the altar
was set out with drawn-work altar-cloths and
clusters of wild lilies in the ramada, that long
two-walled house of wattled brush that served
the Indian so well. Once there was a cloudburst in the cañon behind the rancheria and
the water came roaring against the huts, and
the ramada — but one must really make an
end of incident, and follow after the sheep.

You should have seen Don José Jesús let-
ting his cigarette die out between his fingers
as he told the story of his Long Drive, young
vigor and the high, clean color of romance
lightening the becoming portliness of middle
years. Even then you would miss something
in not being able to pronounce his name with

its proper soft elisions and insistent rhythm, José Jesús Lopez.

Señor Lopez began to be major-domo of the sheep at Tejon in '74, shaped to his work by much experience in the Southwest. In '79, that year of doubtful issues, he left La Liebre on the desert side to drive ten thousand sheep to Cheyenne. He had with him twelve men, none too well seasoned to the work, and a son of the only Henry Ward Beecher for his book-keeper. How this came about, and why Beecher left them before accomplishing the adventure, does not belong in this story, but there is no doubt Don José Jesús proved himself the better man.

They went out, I say, by La Liebre, northward across the Antelope valley when the *chili-cojote* was in bloom and began to traverse the Mojave desert. Well I know that country! A huge fawn-colored hollow, drawn on its borders into puckery hills, guttered where they run together by fierce, infrequent rains; mountains rear on its horizons out of tremulous deeps of air, with mile-long beds of lava simulating cloud shadows on their streaked sides.

Don Jesús went with his sheep in parted bands like Jacob taking out his flocks from Padanaram, dry camp upon dry camp, one day like to every other. If they saw any human traces on that journey it might have been the Owens Valley stage whirling on the thin, hard road, or the twenty-mule ore wagons creaking in from the plain of Salt Wells, stretching far and flat.

All trails through that country run together in the gorge of Little Lake, untwining on their separate errands as they open out toward Coso. Don José kept on northward until he had brought the ten thousand to pasture in the river bottom below Lone Pine, where the scar of the earthquake drop was still red and raw. Enough Spanish Californians had been drawn into that country by Cerro Gordo and neighboring mines to make entertainment for so personable a young man as Don José Jesús, dancing in the patios at moonrise with the señoritas and drinking their own vintages with courteous dons. The flock rested hereabout some weeks and passed up the east side of the valley loiteringly, finally crossing through the White Mountains to Deep Springs Valley, thus far with no ill fortune. That was more than could be laid to most adventurers into that region. A little before that time John Barker had foraged as far north with twenty-two thousand sheep, retiring disgustedly with nine thousand. Said he, " Where we camped we left the ground kicking with dying sheep."

This was the time of the great drouth, when season after season the rains delayed, flinging themselves at last in wasteful fury on a baked, impervious soil. Rack-boned cattle died in the trails with their heads toward the place of springs, and thousands of flocks rotted in the dry ravines. Lopez took his sheep by the old Emigrant Trail, southward of the peak I watch daily, lifted clear white and shining above the summer haze, and came into the end of Deep Springs. The feed of that country is bunch grass with stubby shrubs, shoulder high to the sheep. The ten thousand passed here and reached Piper's in good condition, having drunk last in Owens Valley. Piper was a notable cattleman of those parts, annexing as much range as could be grazed over from the oasis where his ranch house stood, and looked with the born distrust of the cowman on the sheepherder. Notwithstanding, the manners of Don José won him permission to keep the sheep along the stream-side until they should have their fill of water. But sheep are fastidious drinkers, and the water of Piper's Creek was not to their liking.

Now observe, the flock had come over a mountain range and across a considerable stretch of sandy and alkali-impregnated soil since last watering, but they would not drink. Lopez hoped for a living stream at Pigeon Springs, but here the drouth that fevered all the land had left a caked and drying hole. Now they pushed the fagged and footsore sheep toward Lida Valley, where there was a reservoir dammed up for a mine, for there is gold in that country and silver ore, very precious; but an imp of contrariety had been before them, and though the sheep were pushed into it and swam about in the pool sullenly, they would not drink.

All that country was strange to Don José Jesús, bewildering whitey-brown flanks of hill

and involved high mesas faced by dull blue
mountain ridges exactly like all other dull
blue ridges. A prospector, drifted in from the
outlying camps, reported abundance of feed
and water at a place called Stonewall. Lopez
sent men forward with picks and shovels to
make a drinking-place while he came on slowly
with the flock, but after two days he met his
men returning. No water, said they, but a
slow dribble from the cracks of seepage in the
stone wall. Now they turned the flock aside
toward Stone Cabin, footsore, with heaving
flanks and shrunken bellies. At home, they
might feed a winter long on the rain-bedewed
tall pastures without drink, but here on the
desert where the heat and dryness crumple
men like grass in a furnace, the sheep, though
traveling by night, suffered incredibly. All
through the dark they steered a course by the
stars that swung so low and white in the desert
air; morning and evening they fed as they
might on the dry sapless shrubs, and at noon
milled together on the sand. Each seeking
protection for its head under the body of another, they piled hot and close and perished

upon their feet. Made senseless by heat and thirst, they strayed from the trail-weary herders.

Lopez, following such a band of estrays into the fawn and amethyst distances, at the end of two days had lost all his water, and persisting to the end of the third day, began to fail. His men, not finding him where he had appointed a meeting, returned to his point of starting and took up the clue of his tracks; following until they saw him through a field-glass, at last, going forward dizzily in the bluish light of dawn. They had no more than come up with him, when at the relieving touch of water in his parched throat, he fell away into a deep swoon of exhaustion. For three hours his spirit ebbed and tugged in the spent body while the men sheltered him in their own shadows from the sun and waited, as they of the desert know how to wait its processes and occasions. At last, having eaten and drunk again, he was able to make the remaining thirty miles to camp and bring in his sheep to Stone Cabin, where there was a well of fresh, sweet drink. They had come a hundred and thirty miles with the flock all waterless; and Don José Jesús laughed when he told it. He had companioned with thirst; failure had stalked him in the bitter dust; he had seen death camping on his trail; and after six and twenty years he laughed, a little as a woman laughs for remembered love. By which I take it, he is a man to whom the taste of work is good.

The flock drifted northward across Nevada until they came to where sixty feet of Snake River roared in the way. Indian agents, it seems, exist merely to fill agencies. At any rate, the one in charge of the Bannock Reservation would mediate neither for Señor Lopez nor the Indians.

"Any way you fix it, if you get into trouble," said the agent, "don't look to me."

Lopez set a guard about his horses and his camp, sought for El Capitan, and dealt with him as man to man. Twenty-four hours to go through on his feet with his sheep, his wagon, and his men; ten Indians to be paid in silver to aid at the river ford; that was the bargain he made with the chief of the Bannocks. Judge then his consternation as he came to the river border in the morning with the last of his bands, to find three hundred braves in possession of the camp. They ate everything in sight with the greatest cheerfulness.

But El Capitan reassured him. " You pay only for ten."

When there was plainly no more to be eaten, the chief laid the hollow of his hand to his mouth and lifted a long cry like a wolf's howl. Instantly three hundred braves had stripped and plunged into the icy swell of the ford. The chuckle of their laughter was louder than the  rush of its waters. Shouting, they drew into two lines, beating the water with their hands. When the herders brought up the sheep, one and another of them was plunged into the living chute. As they struck the water they were shot forward by long arms; the shoulder of one sheep crowded the rump of another. *Spat! Spat!* went the vigorous, brown arms. The swish of the river, cloven by the stream of sheep, was like the rip of water in closed sluices. The wall of shining bodies swayed with the current and withstood it.

" As I live by bread," says Don José Jesús, " ten thousand sheep went over in half an hour."

The herders, swimming over, formed the dripping flocks into bands, and pushed them forward, for the point where the play of savages turns to plundering is easily passed. Lopez called up El Capitan, and the chief called up the ten. Two dollars and a half of silver money went to the chief, and one dollar and a half to each of his men. The rest of the two hundred and ninety naked Bannocks, having swum the wagons over, played on unconcernedly as boys in the freezing river. Within less than their allotted twenty-four hours, Lopez was clear of the reservation. Some stragglers still stuck to his trail, bent on thieving, and one, professing himself son of the chief, rode after them threateningly, demanding a toll, but was appeased with two dollars in silver, and the flock turned eastward across the tablelands.

All this Iliad of adventure leads merely to the transfer of the flock by sale at Cheyenne— squalid and inadequate conclusion! No, but these are the processes by which the green bough of the man-strain renews itself in the suffocating growth of trade. Not that you should have mutton, but that nature should have men. It was so she put the stamp of efficiency on Señor Lopez, who is now at Tejon as major-domo of the cattle. There have been no sheep on the ranch for some years except the few fat muttons that ruminate under the palms, as effectively decorative in their way as the peacocks trailing hundred-eyed plumage on the green and golden grass, lineal descendants of the fowl that Jimmy Rosemeyre brought across the plains at the tail-board of an emigrant wagon in '54.

If you ask me at a distance from its mirage-haunted borders, I should be obliged to depreciate the holding by one man of so large and profitable a demesne as the Ranchos Tejon, Castac, La Liebre, Los Alamos y Agua Caliente, but once inside the territory of the badger I basely desert from this high position, frankly

glad of so wide a reach of hills where mists of
grey tradition deepen to romance, where no axe
is laid wantonly to the root of any tree, and no
wild thing gives up its life except in penalty
for depredation. Most glad I am of the blue
lakes of uncropped lupines, of the wild tangle
of the odorous vines, of the unshorn watershed; glad of certain clear spaces where, when
the moon is full and a light wind ruffles all the
leaves, soft-stepping deer troop through the
thickets of the trees.

XIII

THE SHADE OF THE ARROWS

CHAPTER XIII

THE SHADE OF THE ARROWS

THERE is a saying of the Paiutes that no man should go far in the desert who cannot sleep in the shade of his arrows, but one must know the desert as well as Paiutes to understand it. In all that country east and south from Winnedumah, moon-white and misty blue, burnt red and fading ochre, naked to the sky, it is possible for a man to travel far without suffering much if only he keeps his head in cover; two hands' breadth of shadow between him and the smiting sun or the hot, staring moon. So if he has a good quiver full of feathered arrows, reedy shafts with the blood drain smoothly cut, winged with three slips of eagle feathers, he sticks them in the sand by their points, cloudy points of obsidian flaked at the edges, and lies down with his head in the shadow. This much

is mere hunter's craft, but the saying goes deeper.

When Indian George had shot Poco Bill, who had "coyoted" his children and caused them to die, — for Bill was a "coyote doctor" who bore grudges against the campoodie, — so that when, by reason of his evil medicine-making, four of George's children had been buried with beads and burnings of baskets, to save the other two George shot him, and when I had offered to go his bail, because it is always perfectly safe to go bail for an Indian, and because I would have be-

haved as George behaved if I had believed as
he believed, Indian George for a thank-offering
brought me treasures of the lore of his clan,
and explained, among other things, that saying
about the shade of the arrows.

Now, when a man goes from his own hunt-
ing-ground, which is the forty or fifty mile ra-
dius from his wickiup, into the big wilderness,
it is to meet perils of many things, against
which, if he carries it not in himself, there is
no defense; against death and perversions and
terrors of madness, the shade of his arrows.
And when it comes to formulating the sense
of man's relations to all outdoors, depend upon
it the Indians have been before you.

There is no predicating what the life of the
Wild does to a man until you know what
arrows he interposes between himself and its
influences. There is much in the nature of the
business that brings him to it, modifying the
play of the wilderness on man; cowboy shepherds and forest rangers, whose work is service and concerned with the moods of the land,
reacting from it not in the same case as the
solitary prospector, the pocket hunter, the her-

mit, the merely hired herder. Every year when
the cattle are driven up from the ranches to
the mountain meadows, the men return from
that venture handsomer, notwithstanding the
tan and the three weeks' beard, than when they
set out upon it; and in the beginning of the
forestry service, when one and another of the
villagers had a try at it before the work sorted
them and selected, one could see how in a season it cleared the eyes and tightened the slack
corners of the mouth. Though they had not
before been tolerable, at the end of that time
they would be worth talking to.

But over the faces of the men whose life is
out of doors, yet to whom the surface of the
earth is merely the distance between places,
comes the curious expression which is chiefly
the want of all expressiveness. They are wise
only in the most obvious, the number of hours
between water-holes, the forkings of the trail,
the points for replenishing supplies; but of all
that vitalizes, fructifies, empty, empty! It is as
if one saw the tawny land above them couched,
lion-natured, lapping, lapping, — it is common
to say in the vernacular of these detached indi-

viduals that they are "cracked," which is a way of intimating that all the sap of human nature has leaked out of them.

These little towns of Inyo sit, as it were, at the gates of the Wild, where seeing men go in and out, going all very much of a sameness, and returning sorted and stamped with the sign of the wilderness; it appears that chiefest of the arrows of protection is a sense of natural beauty. Those who cannot answer to the stimulus of color and form and atmosphere and suggestions of tenderness in the vales and moving strength of mountains, are so much at the mercy of mere bigness and blind power and terrible isolation that it seems all graces wither and die in them. Men of this stamp are curiously prone to stop the vacancies of nature with strong drink, as if somehow they missed the prick of growing and productive fancy. Almost any day you might see one such as this shouldering the door-posts of the Last Chance saloon, or drooped above the bar of the Lone Pine.

But shepherding being a responsible employment, it is evident that if men so unde-

fended went about it they would soon be
weeded out by its natural demands. Be sure,
then, that the vacant type will not often be
found about sheep camps, except it be an occa-
sional hired herder related to his work by
necessity. Every shepherd will have something
worth while in him, though when you talk together, since one of you speaks a tongue not
his own, it does not follow that you may draw
it out. Besides it really is not exigent to a
sense of natural beauty to be able to talk about
it. As if without loquaciousness it were impossible for a man's food to nourish him, or medicine do him good. When one premises an
appreciation of the aspect of the land beyond
the question of its service, it is not invariably
because the shepherd has said so, but because
he exhibits its natural reactions. Should he
lack the chiefest arrow, then the Wild sucks
out of him, along with the habit of ready
speech, most of the fitnesses for social living.
Quickliest you get at the evidence of it by ob-
serving if the man has no shyness in his soul,
but only in his demeanor; whether he exhibits
toward you the avoidance of the rabbit, or with

an untroubled bearing eludes you in his thought. I am convinced, though, that it is not entirely the inconsequence of other people's affairs that clips the speech of the outliers, but the faculty of knowing with the fewest possible hints what the other is driving at. Two Indians, two shepherds, understand each other as readily as coyotes when they cut out lambs from the flock; so, also, my friend and I; but I never know what a sheepherder is thinking about unless I ask him, and not always then.

Most frequently he is not thinking of his troubles, for the lesson most completely learned by the outlier is the naturalness of disaster. It is beginning to be believed by a hill-subduing, river-taming people that trouble also is amenable to the hand of man. But the outlier does not so understand it. He begins by finding the weather beyond his province, and ends by determining death and catastrophe, the shuddering avalanche, the cloudburst, the pestilence, so much too big for him as not to be worth fretting about. As well disturb one's self at the recurrent flux of night and day. If the waters of a dry creek arise in the night, being vexed

at their source by furious rains, as they did in
Tecuya, and wipe out three or four hundred of
a flock, if they are scourged by the hot dust-
blind winds past the herder's power to gather
them up, being a Frenchman he might be seen
to weep, but is not embittered, and begins again.
And when you ask him how he fares, will not
remember to mention such as this without
being asked.

It is said by the casual excursionist into the
outdoor life, and said so often that most believe it, that it destroys caste by obliterating
the differences of men; but in fact the wilderness fixes it by rendering their distinctions
natural. For the Wild has not much power to
suggest the human relation. Social imaginings
are the product of the house-habit and social
use. Much of our interest in other humans
arises in the community-bred necessity of ef-
fecting an adjustment toward them, and to
adjust successfully, needing to know whence
they are derived and how related to other men.
But the life of Outdoors rendering such adjustments superfluous, it is possible to meet

another outlier without prefiguring any relation toward him, and therefore without curiosity.

There is something more than poetry — I do not know just what it is, but certainly not poetry — in the acknowledgment of the power of the Wild to effect a social divorcement without sensible dislocation, though one becomes aware of it only on returning to close communities to discover a numbness in the faculty of quick and multifarious social adjustments. Much of coldness, shyness, dullness, pride, imputed to those newly drawn from the wilderness is in fact sheer inability to entertain relations to incalculable numbers of folk. The relations of the outlier to all other men are of as much simplicity as of one wild species to another; liaisons, conspiracies, feuds they keep locked within their order.

Once when I had a meal with a herder of Soldumbehry's, I had left my cup with him by inadvertence, a cheap, collapsible cup which I was used to carry on the range, and thought not worth going back for. The herder put up the cup in his cayaques; and drifted along the foothills out of my range. Three months later, not having met with me and about to pass through the mountains to the east side, he gave the cup to his brother who held a bunch of wethers fattening for the local market. This one kept it until, at the beginning of the fall returning, he passed it to a herder of Louis Olcese, a scared, bushy-bearded man, like an owl looking out of the rabbit-brush, traveling my way. By the ford of Oak Creek he transferred the cup to his " boss." Him I met on the county road trundling south in his supply wagon. The boss dug up a roll of bedding, untied it, unwrapped a blue denim blouse, unfolded a red bandanna handkerchief, and with this account of it, handed me up my cup. It was worth perhaps a quarter, and any one of these men would have stolen feed from his own brother; but they touched society at no points not affected by sheep. And when you think of it, no one ever heard of a sheepherder " shooting up the town."

Nothing contributes more to the sense of human inconsequence than the unhoused nights of shepherding. In the man-infested places the cessation of laborious noises, the subdued hum of domesticity, give a sense of pause, a hint of dominance, as if we had called up the night in the manner of a perfect servant with sleep upon her arm. But in the Wild the night moves forward at an impulse flowing from unknowable control. Darkness comes out of the ground and wells up to the cañon rims, light still diffusing through the upper sky, a world of light beyond our world. Few things beside man suffer a check in their affairs. The wind treads about the forest litter on errands of its own; you hear it but the more plainly as if blackness were a little less resistant to sound. The roar of the stream rises; even by the gibbous moon it finds the lowest ground. Plants give off insistent odors, have all their power to poison, prick, and tear. A match struck at any hour of the night shows you the little ants running up and down the pine-boles at the head of your bed regardless of the dark, for the night is not an occasion, merely an incident.

Moonlight approaching picks out certain high patches of snow, filtering through unsuspected yawnings of the peaks. Among the high close pinnacles it halts and fumbles, glints like a hard bright jewel along the pillared rocks. At moonrise the shadows of the hills are inconceivably deep, the shade of the pine-trees blacker than the pines. The lakes glimmer palely between them with the pellucid blackness of volcanic glass, reflecting the half-lighted steep, the hollow firmament of stars. Over the rim of them one seems to plunge into the clear obscure of space. By like imperceptible lapses, night clarifies to day. Blackness withdrawing from the sky is reabsorbed by the mountains which show darkling for a time, revealing slow contours as the shadows sink in and in. They collect in lakes and pools in the troughs of the cañons and are gathered to the pines.

The appreciation of this large process, going on independently of the convenience and the powers of man, impinges on the dullest sense, provided only it has a little window where the knowledge of beauty may come in. Its ultimate function is to lap the outlier in an isolation like to that which separates brute species from brute species. It is appreciably of a greater degree in those who sleep always in the open than in the hill frequenters who roof themselves o' nights. You come to the camp of an outlier and are welcome to his food and his fire, but are no nearer to him than a bird and a squirrel grow akin by hopping on the same bough. He accepts you not because you are on the same footing, but because you are so essentially differentiated there is no use talking about it.

"And do you," inquires the community-bred, "go about alone, unhurt and unoffended in the Wild?" What else? The divination of natural caste is extraordinarily swift and keen in the outlier, keen as the weather sense in cattle. Their women-folk, being house-inhabiting, might assume a groundless intimacy, premise a community of interests when necessarily barred from whole blocks of your experience, even annoy by a baseless conceit of advantage, but cowboys and shepherds, trappers and forest rangers, make no such mistakes.

It is true that one carries that in one's belt

to prevent offense at a dozen yards; such as this are the teeth and claws which every inhabitant of the Wild has a right to, and on the mere evidence of carrying about, avoids the necessity of using. But the real arrow of defense is the preoccupation of the motive, the natural and ineradicable difference of kind. It is not in fact the dread of beasts nor the fear of man that causes one to go softly in the Wild, but the assault it makes on the spirit. Knowing all that the land does to humans, one would go fearsomely except that the chiefest of its operations is to rob one finally of all fear,—and besides, I have always had arrows enough.

AFTERWORD

BARNEY NELSON

One of the last pieces of writing Mary Austin produced before her death in 1934 was a poem entitled "When I Am Dead." In the poem Austin says her idea of the perfect afterlife would be to rejoin the shepherds and once again walk behind a flock of sheep through a "saffron-shod evening." She names again many of the shepherds who appear in *The Flock*, Elcheverray, Little Pete, Narcisse Julienne, and says:

So it shall be when Balzar the Basque
And the three Manxmen
And Pete Giraud and my happy ghost
Walk with the flocks again.

The time Austin spent following along with the sheep and interviewing the herders seems to be among her happiest and most insightful moments. Periodically, the theme of returning to those days reappears briefly in her work. In a poem called "The Passing of the Flocks" she says:

I should like to be in Inyo
When the flocks go by again,
The long-fleeced flocks and the wise old dogs,
And the long-armed bearded men. (AU 436)

An unpublished scrap of a manuscript, which was probably the beginning of a new book, can be found among the Austin papers at The Huntington Library. The working title hints at the scope of her plan: "The Ancestry of the Modern Flock, the Beginnings of the Weaving Craft, Weaving Frames, the True Loom, Batten, and Shuttle, the Navaho Loom, the Preparation of Wool, Carding, Spinning, and Dyeing, the Development of Dyeing" (AU 15). Never before and perhaps never again does Austin's life achieve the same level of peace and close connections to nature as she enjoyed during the years when she lived close to sheep and shepherds. Several critics say her writing never again reaches the quality she produced in her earliest books.[1]

Austin was born in Carlinville, Illinois, in 1868. Her childhood and young adulthood were marred by the death of both her father and a younger sister. In 1888 Austin, along with her mother, older brother Jim, and younger brother George, left Illinois to homestead in the Southern California mountains. Austin published a brief memoir of their journey, "One Hundred Miles Horseback," in Blackburn College's student literary magazine. She had just graduated with a degree in biology from Blackburn.

Shortly after the family's arrival in California, Austin met and became friends with General Edward Fitzgerald Beale, the owner of the Tejon Ranch, a huge old Spanish land grant that sprawled across the meadows and foothills of the southern San Joaquin Valley. General Beale was flattered by Austin's sincere interest in the sheep culture, and he often rode with her horseback across the pastures, explaining the history and lore of the California shepherd.

During this period, Austin kept a scrapbook/

notebook, which she filled with observations like: "An eagle swooped down and carried off a lamb not 30 yards from us and rose slowly. Jim shot it. It measured seven feet seven inches from tip to tip of wings and four feet four inches from beak to tail. The lamb did not make any noise when the eagle struck but the eagle screamed harshly when the herder struck at it with his staff" (AU 267, 2).[2] She gathered memorabilia and photographs and made sketches from them that would enhance her descriptive powers as well as someday illustrate her publications. The photographs, drawings, and memorabilia in the scrapbook/notebook often include labels in Austin's handwriting: "Noriega's camp," "flocks coming in to shearing," "in the sheds," or simply "the crook," "sacking frames," or "sheds."[3] One of the scraps of memorabilia collected is the badger logo and letterhead for Beale's "Tejon Ranchos." Austin evidently drew a sketch of the badger and sent it, along with others, to E. Boyd Smith, the official artist for *The Flock*. A professional, but very similar, version appears on the chapter opening for "Ranchos Tejon."

Art is an aspect of Austin's lifework that has heretofore gone unsung by scholars. Hidden within the many folders preserved in The Huntington's collection are several fine line drawings and watercolors by Austin. She evidently sent *The Flock*'s New York publisher such well-sketched drawings of wildflowers, animals, shepherd carvings, and equipment that the official artist seems to have copied her work almost line for line. Drawings for chapter openings such as "The Hireling Shepherd" and drawings scattered throughout the book, such as the locoweed that appears on page 47 and again on 125, were obviously redrawn from Austin's sketches.

In a set of sketches of bells and bell leathers, her drawings include interesting notes about characters who appear in *The Flock*. She says that one bell and yoke belonged to "Narcisse Duplin's best-leader called Le Diable." The artist's drawing of the same bell opens the chapter "The Open Range." She worked hard to

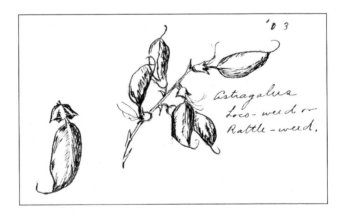

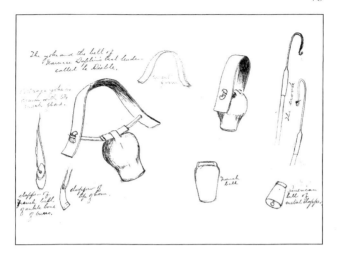

give the artist authentic detail from which to work, comparing French bells and American bells or explaining that "Pete says yoke is drawn with too much flare." Her notes indicate that one clapper is for a French bell and made from the "ankle bone of burro" and that another is a "clapper of tip of horn."

Austin often turned photographs collected in the notebook into drawings, such as the following pair of a wool sacking tower, which again influenced *The Flock*'s artist for his drawing on page 40. Although many of the photographs are

obviously snapshots and poorly exposed, one stunning photograph of "Little Pete's band passing on the Long Trail south of Lone Pine," included in her collection and labeled with her handwriting, was obviously done by a professional photographer working with a large-format camera. The wonderful chiaroscuro layering of snowcapped Sierra and black foothills strongly resembles a photograph made several years after

Austin's death by Ansel Adams entitled "Winter Sunrise, The Sierra Nevada, From Lone Pine, California, 1944." Austin was instrumental in helping Adams begin his photography career when he provided the photographs to her text for *Taos Pueblo* (1930). She was at the height of her fame, while Adams was a young photographer on assignment for his first book. Later he also illustrated a 1950 edition of Austin's *The Land of Little Rain* that included numerous photographs of

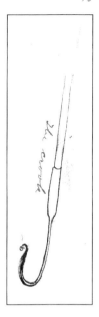

sheep and shepherds taken in and around Austin's shepherd trails. Austin's art and her influence on artists should be further pursued by scholars.

During what proved to be an unhappy marriage to Wallace Safford Austin, a self-styled entrepreneur who dabbled in prospecting, irrigation schemes, school teaching, and land office work, Mary often found herself financially destitute and alone. At one low point she was evicted from their rooming house while he was off prospecting. Humiliated and sitting on a trunk containing their belongings, which had also been placed in the street, she wondered what to do. Desperate, she found a job baking pies and cooking at a boardinghouse "on the far edge of town" and was able to feed herself and her husband when he came home that evening (*Earth* 235). Eventually, she turned to writing as both a way to make money and as an outlet for her emotions.

For many years, her personal life just seemed to get worse. In 1892 Austin's daughter, Ruth, was born retarded and often spent her days and nights screaming. Needing relief, Austin asked for help from neighbors and even her mother. Few could stand the situation for more than an hour, yet they were quick to condemn Austin's failure to be patient with and dedicated to her child. Neighbors gossiped that Austin would tie her daughter in a chair and then lose herself in her writing, concentrating on finding just the right word to describe a mountain while Ruth screamed. Austin's mother blamed her daughter for producing a retarded child and said, "I don't know what you've done, daughter, to have such a judgment upon you." Austin responds simply that "It was the last word that passed between us" (*Earth* 257). Thus Austin's writing became not only a means of support but also an escape from the torment of a screaming daughter and an absent and nonsupportive husband.

During these years, one highlight of Austin's existence was the passing of the flocks on a trail close to her home. Turning to the shepherds for friendship, she would often dash out to the trail to take vegetables from her garden to trade for a piece of mutton or, after Ruth was asleep, hike out to the blinking shepherd campfires to listen to stories. Shepherds and sheep ranchers— whether Hispanic, Basque, Paiute, or French— became her closest friends and supporters. So it is not surprising that Austin began to use her pen to defend this traditionally maligned group. Her first nationally published stories were positive characterizations of sheep and shepherds: "A Shepherd of the Sierras" (*Atlantic Monthly*, July 1900), "The Little Coyote" (*Atlantic Monthly*, Feb. 1902), "The Last Antelope" (*Atlantic Monthly*,

July 1903). Her first novel, *Isidro* (1904), had a shepherdess heroine.

Around the turn of the century, just before Mary Austin wrote *The Flock* (1906), Los Angeles began to build the aqueduct that would eventually spirit the eastern Sierra's abundant water away from her Owens Valley neighbors. At the same time, railroads also began to promote national parks, which would eventually close thousands of acres of high Sierra meadows to summer sheep grazing. These two major blows to irrigating and grazing would put many of her friends out of business. Austin had probably also just read Frank Norris's *The Octopus* (1901), a novel suggesting that California and the nation were at the mercy of a huge, ruthless, and manipulative railroad monopoly. As Norris's novel opens, a band of sheep have accidentally wandered onto a railroad right-of-way in the path of a fast-moving train. He describes the scene as:

a slaughter, a massacre of innocents. The iron monster had charged full into the midst, merciless, inexorable. To the right and left, all the width of the

right of way, the little bodies had been flung; backs
were snapped against the fence posts; brains
knocked out. Caught in the barbs of the wire,
wedged in, the bodies hung suspended. Underfoot
. . . the black blood, winking in the starlight, seeped
down into the clinkers. (Norris 42)

As suggested by Norris's images of sheep as
innocent victims of the thundering machines of
big business, much of the rhetoric and symbolism regarding private and public land, rural and
urban conflict, tourism and grazing, water rights
and railroads, socialism, capitalism, and democ-
racy centered around sheep.

As regional, indigenous people of various
races were pushed aside for the "greater public
good," which almost always turned out to be
oppressive, urban, and in some way lucrative
for those in power, Austin rose to defend her
friends and neighbors. In *The Flock* she blends
natural history, politics, and allegory into a
genre-blurring narrative championing local shepherds in their losing battle against the quickly
developing tourist business. Using a complex
essay pattern that Carl Bredahl calls "divided

narrative" (49), Austin structures *The Flock* as a
loose but interwoven collection of sketches about
sheepherding in the Sierra, merging nonfiction
and myth. Her multivoiced narrative style, later
of much interest to class and postcolonial theo-
rists like Mikhail M. Bakhtin and Trinh T.
Minh-ha and probably inspired by Austin's
attention to Native American storytelling,
enabled her to use the oral tradition in a written form. Carefully citing her sources in the first
chapter, Austin gives multivoiced credit to historic California sheepherders, from Hispanic
landowners to the "Basco" shepherds:

I suppose of all the people who are concerned with
the making of a true book, the one who puts it to
the pen has the least to do with it. This is the book
of Jimmy Rosemeyre and José Jesús Lopez, of Little
Pete, . . . of Noriega, of Sanger and the Manxman
and Narcisse Duplin, and many others. (*Flock* 11)

With this introduction, Austin creates for herself a humble journalistic persona in order to
give author(ity) to the common working-class
shepherds, rather than to herself as an all-know-

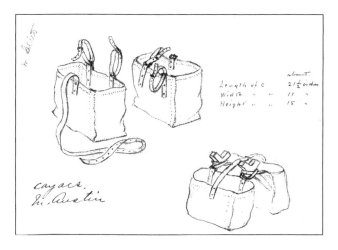

ing author. Following a detailed history of the
sheep industry in California, Austin romanticizes
shepherd work in the Sierra with lyrical and
allegoric vignettes about sheep management,
shearing, shepherd ways, grazing the Sierra,
weather sense, flock mentality and habits,
sheepdogs, wars with cattlemen, predators, the
Tejon Ranch, and the coming of a national
park. This meticulous attention to detail gives
Austin's work an unmistakable sense of veri-
similitude. One example of this attention to
detail appears again when comparing the artist's

drawings to Austin's. On page 58 appears the artist's rendition of her drawing of "cayacs." Austin's version includes such minute detail as stitching and punched holes, and her hand-noted measurements indicate length, width, and height. This is also one of the few drawings that Austin actually signed as the artist with "M. Austin."

In one chapter, "The Sheep and the Reserves," Austin blatantly argues against the call to ban grazing within Yosemite National Park for the sake of developing tourism, declaring that even the rangers often sympathized with law-breaking shepherds and looked the other way while they sneaked back into their old summer haunts. The rangers, she claims, "despised . . . the work of warding sheep off the grass in order that silly tourists might wonder at the meadows full of bloom" (*Flock* 193). Austin obviously wrote *The Flock* as a response to political activism and publicity promoting tourism and national parks, as well as the underlying economic and cultural class divisions that such activism encouraged. In her final chapter, "The Shade of the Arrows," Austin uses a Paiute saying, "no man should go far in the desert who cannot sleep in the shade of his arrows," to give warning about visiting a "wild" place when one does not truly understand how to survive there (253). This saying seems to be the theme of the entire book: that working shepherds belong in the Sierra and tourists do not.

Wherever she found it, Austin challenged the stereotype of rural people as ignorant and illiterate. She recognized art in humble places, granting sophistication to basketmakers and pocket hunters, and remembering that she never found a better companion with whom to discuss French literature than a dark shepherd she calls "Little Pete." She respects and recognizes shepherds as creative: poets, wood carvers, musicians, storytellers, and philosophers. Both her writing and her collected papers contain extensive examples of Austin's respect for the artistry of the shepherds. One stanza of her poem "The Passing of the Flocks" talks about the shepherd as storyteller:

Then a herder will tell you a story
Of a mountain lion he slew,
And how he tracked it home to its lair,
Till he makes you wish it were you,
And one will tell of a bear in the night,
And one has an eaglet tame
Which he took from a dizzying eyrie
On a peak without a name. (AU 436)

Her drawings of the carvings on bell-leather
keys and the "staff of a Basque herder ½ size"
are examples of the art she often found in their
humble camps. Austin's sketches again inspired
The Flock's artist for drawings on pages 103 and
105.

Austin believed the perception that shepherds
were ignorant often stemmed simply from language barriers and writes defiantly that "these
Bascos are a little proud of the foolish gaspings
and gutterings by which they prevent an understanding" (*Flock* 60, 63). Austin's defense of
sheepherders begins her lifelong pattern of
respectful representation of indigenous, regional
cultures. She often preferred the company of
Shoshone, Paiute, and settlers from Old Mexico

to that of her white neighbors whose creeds, she
points out, were "chiefly restrictions against
other people's way of life" (*Land of Little Rain*
106). In *The Flock*, Austin pointedly explains
that management of a flock was "never a 'white
man's job,'" and that a white man hired in that
position was one of "the
impossibles" (62). Rather
than the typical attitude
that a "white" man was
"too good" for the job,
Austin implies that he
was not good enough. A
sheep owner who would
hire unskilled white men
to take care of his flock,
she says, brands both the owner and the
hirelings as incompetent—boomers after a fast
buck, not sheep people. Real shepherds, as
Austin carefully explains, usually began as children and were taught by parents who came from
a long line of shepherds (*Flock* 51–54).

One poem, "The Shepherd Wind," collected
in Austin's book of children's poetry, *The
Children Sing in the Far West* (1928), attempts

to pass on a pride in this heritage to shepherd
children who were her pupils at Lone Pine:

The wind is the shepherd who drives the clouds
    Across the field of sky
And fast or slow, as the wind may blow,
    I watch his flocks go by.

His folds are under the rim of the world
    North, south, and east and west—
And the shepherd cares for all of his sheep
    But he loves the lambs the best.

In clear, smooth paths of the middle sky,
    And highest up in the air
He lets the snow white cloudlings run
    Whenever the days are fair.

The mother clouds on the mountain browse,
    Or nose along the hill,
Or sleep in the shade of the tallest peaks
    When the shepherd wind is still.

The cloudkins come at his lightest call,
    And he whistles the mothers home,

But he shouts aloud at the black storm cloud,
    For black sheep love to roam.

If ever the tale of the flocks is short,
    Or the shepherd miss his sheep,
He calls by the folds where the white frogs lie
    And whips them out of the deep.

Last night when the gates of the west were ajar
    To let the sun slip through,
I saw the little clouds scampering far
    Into the open blue.

And every one had a golden fleece,
    But a few turned rosy red
As they huddled down the coast-wise hills
    Where the willful black sheep led.

While all night seeking his truant sheep
    The whistling wind went by,
He found them at last on the edge of dawn
    Under the lee of the sky;

He drove them back by the shouldering hills
    That fence the great sea lane,

I heard the beat of their scampering feet,
But others thought it was rain.

And we can follow the way he took
To the pastures of the deep
For every hill has a snowy fleece
It stole from the truant sheep. (15–17)

She says she wrote the poem forty years prior to its publication while she was teaching school in Lone Pine, because she "felt obliged to have something for my pupils about the land they lived in" (vii).

In *The Flock*, Austin humbly claims no personal experience herding sheep. Instead of presenting herself as an authority, she presents herself as a simple reporter, asking questions, listening to shepherd stories around campfires. At her best, Austin practices the humble techniques of journalism and storytelling, meticulously gathering facts and stories from many different points of view, rather than trusting to her own knowledge. However, Austin actually possessed a profound personal understanding of the rural culture. She had "grown up in a farming country, of farming kin" (*Earth* 227), spent two years homesteading on the borders of California's old Tejon Ranch, and fifteen in the Owens Valley, living for almost two decades among shepherds and beside what they called "The Long Trail," which wound through the Sierra (*Flock* 12). She knew and respected the shepherds she wrote about and tried to represent the culture from the inside.

Although critics consistently list *The Flock* as one of Austin's best books and say her sheep have the "potency of symbol,"[4] in the end the book has usually been read as a charmingly outdated pastoral that inspired the reader to want to "lie under the sky with dogs and flocks, lulled to sleep by the 'blether' of ewes and the bark of distant coyotes" ("A Review" 17–18). By 1930, critics were already calling *The Flock* one of Austin's "least known volumes" (Tracy 24); and one modern critic dismisses it as "a study of the ways in which the insistent claims of motherhood can inhibit one's distinctive voice" (Wyatt 87). Austin, however, is not simply presenting a rhetorical argument about sheep grazing, nor can *The*

Flock be interpreted as a feminist manifesto. Her characterization of working people, her use of both symbolism and fact, her focus on storytelling and its democratic implications, and her creation of a journalistic persona all work together to achieve *The Flock*'s unity as an argument supporting the democratic principles of equality and government by consent of the governed.

Austin's story line in *The Flock* was obviously based on a timely political controversy raging in California at the turn of the twentieth century, but these are also ancient conflicts, and sheep and shepherds carry ancient allegoric meaning. A glowing 1906 review in *The Nation* compared Austin to Virgil, saying,

Badly stated, [*The Flock*] is no more than a study of the sheep industry in California, with a slender thread of historic narrative, a picture of sheep herding, a word for irrigation. This summary of *The Flock*, however, bears about as much relation to the actual achievement as a statement that the first book of the Georgics is a treatise on agriculture. ("A Review" 17)

Arguing on one level against banning shepherds and sheep from Yosemite, at a deeper level she argues against dividing people, animals, or places into hierarchies and classes. She defends indigenous working people and recognizes the importance of ecological interdependence between the place and its inhabitants. Through sheep, both real and allegoric, she champions cultures that remain in close contact with natural rhythms and resources and believes these will prove sustainable. Ecologically, in order to sustain the complicated web we call life, Austin believes plants, soils, insects, water, and animals, whether wild, domestic, or human, all depend upon one another.

Through what she calls the "pale luminosity" of sheep dust rising along the Long Trail, Austin conjures "the social order struggling into shape" (*Flock* 57) and follows the ancient storyteller's tradition of using sheep to represent the human masses. Both she and her readers were well aware of that tradition. With affection, humor, and respect, she describes lambs as they struggle to adapt to the unique regional challenges of California's Long Trail:

Young lambs are principally legs, the connecting body being merely a contrivance for converting milk into more leg, so you understand how it is that they will follow in two days and are able to take the trail in a fortnight, traveling four and five miles a day, falling asleep on their feet, and tottering forward. (*Flock* 25).

Instead of portraying sheep and lambs as helpless female masses who are rapidly breeding more and more ragged children, Austin allegorically characterizes young lambs as the hope of the future. Quickly able to stand on their own feet, she says they totter forward, even when physically exhausted—a fiercely proud attitude toward the nation's "common" youth.

Austin uses accurate and detailed natural history about real sheep to express her respect for the intelligence of both common animals and common people. She finds the natural laws that operate within flocks and between shepherds, dogs, predators, and sheep quite similar to the natural laws that operate within society and between races, classes, and genders. Prejudice, in Austin's mind, is actually ignorance. In *The*

Flock she carefully explains the complexity of sheep social structure as a culture: sheep intelligence, range of emotion, cravings, watch keeping, selection of leaders, and communication patterns. She describes their recognition, exclusion, and finally acceptance of strangers.

Austin's sheep-induced political views were neither socialistic nor hierarchal. Lawrence Clark Powell (1971) calls her a "Fabian Socialist," someone who favors supporting gradual social progress while avoiding direct confrontation with the state. However, in Austin's mind the allegoric "flock" was not a cohesive, socialistic, cooperating group, but a seething stew composed of highly independent individuals. She actually objected to the idea of unity, especially in thought. Allegorically, she says, the flock-mind is less than the sum of all the intelligences of individual sheep (*Flock* 109). Carefully explaining complex social structures, she argues that sheep can think for themselves but simply find life less complicated when they select and follow leaders. She claims that the "earliest important achievement of ovine intelligence is to know whether its own notion or another's is most worth while, and if the other's, which one" (*Flock* 109–10).

The flock-mind, she says, is neither natural nor permanent, but artificial and temporary. She notes that once a sheep is separated from the flock, the shepherd must find it, because it will not return on its own: "it is for them as if the flock had never been." Also, she supposes, sheep may very well understand the arm signals and will of the shepherd, but only grudgingly obey, feeling "a little resentful of the importunity of the dogs" (*Flock* 110, 119, 117, 127). Austin believes that this intellectual wildness will save the world. Like Thoreau, she finds it lurking just below the thin veneer of domesticity and breaking out when oppressed people or animals reach a point of desperation. At the end of *The Flock*, she calls this natural wildness one of the "arrows" that inhabitants of the wild must learn to "sleep in the shade of" in order to survive. Real sheep, Austin suggests, when studied deeply, the way shepherds study them, can help us understand our own political and social problems. Real sheep are not stupid, easily controlled, or of one

mind, although they do have a few idiosyn-
crasies.

One reviewer observes that although Austin
"stops for queer speculations on the development of the animal mind," she "does not senti-
mentalize about them: she makes the limits of
instinct quite as clear as its scope" ("A Review"
17). Carefully presenting the wild habits of
domestic sheep, Austin's allegory ranges from
Thoreau to biblical authors who wrote that the
meek would inherit the earth. Austin does not
consider rural people or sheep as helpless innocents in need of protection from powerful, more
intelligent capitalists. Instead, she stresses that
the "sheepmen had always the advantage in
superior knowledge of the country, of meadows
defended by secret trails and false monuments,
of feeding grounds inaccessible to mounted men,
remote, and undiscovered by any but the sheep"
(*Flock* 198). Austin believed in the struggle and
was not asking to have rural people "protected"
from the captains of industry. She never
wavered in her faith that "common" people can
and should take care of themselves. In direct
contrast to social Darwinism, she did not

believe that the "best people" always rose to the top, either. Her shepherds were in constant flux from the estate of owner to hireling, and this cyclic flux, she says, makes all shepherds philosophers (*Flock* 61).

In the chapter called "The Go-Betweens," Austin's allegory takes on another dimension when she describes the way shepherds have domesticated the predator in order to guard sheep against its own kind. Like ex-criminals who sometimes make the best police because they are not easily fooled, predator types can make a living either by preying on sheep or by protecting them and dining on choice mutton killed, butchered, and served to them in a dish by the shepherd. Exactly who has trained whom is a question Austin poses but doesn't answer. *The Flock* is full of heroic, intelligent dogs: dogs whose loyalty to the shepherd is legendary, who carry on without the shepherd, who kill coyotes without help, who can pick a sheep from the herd when the shepherd speaks only the animal's name. Austin observes that sheep are "silly" regarding dogs and other predators, because they never seem to be able to figure out which predators are their friends and trust them. Yet, when viewed against the complexity of the sheep/dog relationship, this silliness may actually be a form of wisdom, because some sheepdogs will turn on the sheep (*Flock* 135–52).

Austin's complicated view of the predator/prey relationship is also nonjudgmental. She does not consider predators thieves but simply animals who are following their own natural behavior patterns. Quite often she assumes humans operate under similar unique behavior patterns. In the chapter called "The Strife of the Herdsmen," Austin explains how shepherds, under orders from their employers and with their own traditional sense of responsibility toward the welfare of their sheep, are forced to match wits against each other and the forces of God, beast, weather, and park rangers in order to feed their flocks. Desperate shepherds infringed on the forbidden pastures in the national park and, she says, "came out boasting, as elated, as self-congratulatory as if they had merged railroads or performed any of those larger thieveries that constitute a Captain of Industry" (*Flock* 197). She sprinkles her writing with constant reminders that under the right circumstances we are all not only animals but also thieves: "Times when there is moonlight, watery and cold, a long thin howl detaches itself from any throat and welters on the wind" (*Flock* 93).

Austin argues that problems with predators are often caused by close herding and relates an incident at El Tejon, during the drought of 1876, when fifty-eight thousand head of starving sheep were turned loose in December to die. The staggering flocks slowly disappeared into the bear-, cougar-, and wolf-infested mountains. The next fall, when rains finally replenished lowland meadows, fifty-three thousand head of healthy sheep trailed themselves back down for the winter (*Flock* 235). Sheep and predators have shared the same pastures since time began, says Austin, and in her mind, sheep were never huddled, helpless masses, too stupid to take care of themselves. According to Austin, shepherds have no quarrel with predators either:

It is only against man contrivances, such as a wool tariff or a new ruling of the Forestry Bureau, that the herder becomes loquacious. Wildcats, cougars, coyotes, and bears are merely incidents of the day's work, like putting on stiff boots of a cold morning, [or] running out of garlic. (*Flock* 176)

Bears, she muses, often stroll harmlessly over sleeping shepherds at night or burn their paws trying to rob frying pans. "Or so it was," she states pointedly, "in the days before the summer camper found the country" (*Flock* 186).

Realizing how powerfully Austin structured her literary argument supporting working shepherds, John Muir may have felt that both Yosemite National Park and tourism itself were threatened. Muir's book *My First Summer in the Sierra* (1911) may well have been written in reaction to *The Flock* (1906). Muir favored preservation of "pristine wilderness" as a place for leisure and study and revered those parts of the West that most resembled Europe's mountains, waterfalls, and lush woodlands as inspirational to writers and artists. Austin, on the other hand, believed land should be valued as home and argued that ranking land into hierarchies and preserving the most beautiful encouraged abuse of the unbeautiful. On both literal and symbolic levels, Muir and Austin each desired to protect natural resources from ravaging hordes, but their perceptions regarding who those hordes might be differed considerably. Muir feared the common rural masses and their domestic animals. Austin feared the unquenchable urban thirst for water and recreation. Muir wanted to preserve beautiful places for escape and enjoyment. Austin wanted to preserve sustainable rural communities.

On the subjects of sheep, shepherds, and sheepherding, Muir contradicts Austin's descriptions in *The Flock* almost line for line. Austin's sheepherder seldom carries a six-shooter; Muir's usually does (*Flock* 83; *My First* 129). Austin goes into rapture over the wonderful meals she has eaten at a shepherd's fire; Muir says the sheepherder's food is "far from delicate" (*Flock* 82–83; *My First* 81). Austin argues that "[t]he smell of sheep is to the herder as the smack and savor of any man's work"; Muir makes fun of the sheepherder's desire to sleep next to the sheep "as if determined to take ammoniacal snuff all night"

(*Flock* 93; *My First* 129). Austin calls ridicule of shepherds simple prejudice and argues that the fact "[t]hat most sheeperders are foreigners accounts largely for the abomination in which they are held and the prejudice that attaches to the term" (*Flock* 55–56).

Austin also suggests that because shepherds often spoke an(other) language, outsiders came to the "unfounded assumption" that most sheep-herders were "a little insane." She suggests further that their outdoor life "nourish[es] the imagination, and they have in full what we oftenest barely brush wings with" (*Flock* 61–63). Although Muir claims that mountain hardships produce strength and wisdom in himself and fellow mountaineers, he believes this same solitary outdoor life adversely affected the shepherd: "seeing nobody for weeks or months, he finally becomes semi-insane or wholly so" (*My First* 24). Austin states emphatically, "[w]ith all my seeking into desert places there are three things . . . I have not seen,—a man who has rediscovered a lost mine, the heirs of one who died of the bite of a sidewinder, and a shepherd who is insane" (*Flock* 65). Muir just as emphatically states,

"The California shepherd, as far as I've seen or heard, is never quite sane for any considerable time" (*My First* 24). Muir also claims the sheepherder wears "everlasting clothing" consisting of pants waterproofed by drips of "clear fat and gravy juices" that have clustered into stalactites and become imbedded with bits of nature. These greasy clothes do make the sheepherder a collector of specimens, but "far from being a naturalist," jokes Muir derisively, keeping the class and "foreign" divisions clear between himself and the shepherd (*My First* 130). At the level of allegory, a comparison of their work becomes almost ominous. Muir's sheep, representing the common people, are a burden on society, increase the rapid depletion of natural resources, and face a life of increasing poverty.

Austin persistently questions cultural, educational, political, and language hierarchies, while Muir just as persistently characterizes himself as superior to shepherds, because they are, in his view, dirty, uneducated, and racially other. While being out in wild nature inspires Muir to a deep appreciation, the same exposure does not seem to work on his fellow shepherd who, according to Muir, ignores the wild beauty that surrounds him (*My First* 41). Not so, insists Austin. The shepherd simply finds it difficult to put his appreciation into words. She maintains that "it really is not exigent to a sense of natural beauty to be able to talk about it" (*Flock* 258) and believes, as Shakespeare did, that silence is sometimes a sign of such great love that it cannot be expressed in words.

Perhaps partly motivated by his own embarrassing inability to learn shepherding skills and handle what he considered a simple unskilled job, Muir publicly curses sheep for their "reckless ravages" and calls them "ruthless denuders" or "hoofed locusts." Austin points out that California shepherds came from countries where sheep, flowers, steep mountain pastures, and crystal streams had coexisted, and probably coevolved, for thousands of years (*Flock* 32, 52). While Austin admits that "[y]ou will find the proof . . . in the government reports" that sheep grazing can cause damage if shepherds are not given their own "fixed" pastures, she makes clear that the damage should not be blamed on the shepherds, who are just following orders,

nor on the sheep, who like all creatures "use the
face of the earth to better it" (*Flock* 172, 206–
10). "No doubt," she argues, "meadow grasses,
all plants that renew from the root, were meant
for forage" (*Flock* 100). Shepherds and sheep
merely follow the same techniques practiced by
good gardeners when they pinch back, prune,
fertilize, and burn in order to produce more
bloom and healthier plants.[5] She explains that
what can appear to be destruction can actually
be cultivation. Grazing through pastures in early
spring as growth began and back again in the
fall after seeds had dropped, sheep made ideal
gardeners for flowering plants. Defending even
the sheep themselves as laborers, Austin
observes that their prunings and droppings had
been fertilizing and strengthening the beautiful
flowered meadows and asks, "Is it not the custom otherwhere [everywhere] to put sheep on
worn-out lands to renew them?" (*Flock* 170,
208–9). Perhaps as proof of her claims, Austin
produced a small but beautiful collection of delicate watercolor paintings of wildflowers, again
now preserved at The Huntington (AU 763).

Austin also realized that Muir's tendency to

imagine the Sierra as pristine wilderness or a
Garden of Eden ignored centuries of Native
American land management. The open, sunlit,
parklike quality of the Sierra, which so
attracted tourists, had been fire-maintained as
pasture, first by native people, then by shepherds. Ironically, Muir wanted to protect the
Sierra "wilderness" from the very people and
animals who had been maintaining it. By the
time Muir first saw the San Joaquin Valley,
sheep had been grazing it for over one hundred
years, numbering over three hundred thousand

in the Valley in 1833 (*Flock* 7). Yet, thirty-five years later, in 1868, Muir describes this sheep-ravaged valley as "the floweriest piece of world I ever walked."[6]

If a duel over grazing Yosemite was indeed being fought between the two authors, Mary Austin lost the battle.[7] Sheep bells no longer tinkle through high Sierra meadows. Muir's persona "worked" on the reader and Austin's did not. Muir's persona appeals to readers who want to imagine themselves as inherently superior to the common masses. Austin, on the other hand, pooh-poohs any idea of heroic risk, saying city dwellers are often amazed that a person can "go about" these mountains "unhurt and unoffended by the wild" (*Flock* 265). The Sierra, she says, was a maze of sheep trails, footpaths, and shepherd camps. Tucked away at the edge of most large meadows were food, shelter, and firewood. Rural western hospitality traditionally welcomed anyone to those supplies (*Flock* 265). Even the distinguished Joseph LeConte, leading a group of hungry hiking students, once robbed a shepherd of his dinner right off the fire with Muir's help and blessing

(O'Neill 30). The presence of shepherds in the Sierra took any true risk out of traveling light in Muir's day, a fact often overlooked by today's young hikers who succumb to exposure and exhaustion trying to follow in Muir's footsteps, some of which may have been made only by his fictive persona (Bowen 164–65).

Readers who prefer to think of themselves as inherently superior to the "average" citizen are put off by Austin's persona, because it humbles them. She insinuates that at least in the mountain pastures working shepherds possess a certain kind of common sense and wisdom that has not been learned by wealthy, highly educated visitors or housebound readers on vicarious adventures. Austin's persona seems to warn the reader to "Stay away and leave these mountain pastures to the people who have given their lives to them." And keeping people away was probably her intent. Even as a child, when Mary Austin was admonished by her mother not to "antagonize people," but to try to "draw people to you. Mary would reply stubbornly, 'And what would I do with the people after I have drawn them to me?'" (Fink 33).

Ultimately, Muir won the political battle over Yosemite grazing, but perhaps he never considered the cumulative effect of his own behavior: that by drawing people to the mountains through his books, he might have been the "worst enemy the wilderness ever had" (Bowen 163). Today, over two million travelers stand in line to see Yosemite Falls. The "beautiful," well-kept gardens full of flowers, "grass up to a bear's hips," and "champagne water," which Muir encountered during that long lost first summer with the sheep, have been replaced by two hotels, four swimming pools, five grocery and general stores, five souvenir shops, two golf courses, six gas stations, a bank, a hospital, campsites for six thousand people, and a vast motel and parking lot complex (Bowen 166). The meadows are ribboned with deep back-packer trails. Park police wear riot helmets, and campers must carry water purification kits.

The struggle Austin illuminates in *The Flock* is still going on today. "Are You an Environmentalist or Do You Work for a Living?" asks Richard White in the title of a provocative new essay. Commenting on today's version of this argument, White asserts that modern environmentalists, "[h]aving demonized those whose very lives recognize the tangled complexity of a planet on which we kill, destroy and alter as a condition of living and working, . . . can claim an innocence that in the end is merely irresponsibility" (185). White proposes that modern environmentalists, while "celebrating the virtues of play and recreation in nature," often "take one of two equally problematic positions toward work." They either "equate productive work in nature with destruction" or they "ignore the ways that work itself is a means of knowing nature" (171). When society privileges leisure over work, as White points out, "Nature may turn out to look a lot like an organic Disneyland, except it will be harder to park" (185).

Modern versions of this debate often use much the same rhetoric, and domestic animals still allegorically represent these clashing cultures and class values. Ironically, according to Robert Bauer's presentation at a recent John Muir conference, Snelling, the little grazing community where Muir wrote his first journal about sheepherding, had a population of 315 in Muir's day and has a population of 315 today.

How threatening *The Flock* may have been to California or national politics in 1906 is probably impossible to determine today, but Austin's later friend, President Theodore Roosevelt, who developed the national park system, appears to have been enough impressed by her arguments to send "a forestry expert to interview her" (*Earth* 289). When the land belongs to no one, and its welfare is no one's responsibility, then every year, she would have explained, the "best contriver" will possess the best pastures. As a result, permanent federal-grazing leases, a method she recommends in *The Flock* and says will help stop grazing abuses on public land (171–72, 209), were implemented and are still in effect today.

After *The Flock*, Austin's literary success was firmly established, and she was able to institutionalize her daughter and leave her husband. By the time Virginia Woolf published her landmark essay in 1929, saying that a woman needed a room of her own and a pension in order to write, Austin had been supporting herself with her writing for over twenty-five years. She was living in her own rambling Spanish-style home in Santa Fe, decorated with rare pieces of Native American and Hispanic art. She regularly entertained dignitaries from all classes, races, and genders. She had traveled the world and enjoyed personal friendships with film stars, artists, and two U.S. presidents. She had been appointed to several important state and national committees and boards and was in great demand as a public speaker. She had written over 30 books and 250 articles that spanned and invented several genres: fiction, drama, short story, political essay, poetry, literary criticism, history, intellectual prose, divided narrative, third-person autobiography, and more. Not only had she already surpassed Woolf's dreams for the woman writer's future, but also Austin had stood up alone and tried to fight the giants of California politics and business. She lost the battle, but as more and more of her work is reprinted and gains more and more public and academic attention, she may yet win the war.

1. The Mary Hunter Austin Collection at The Huntington Library contains extensive material on *The Flock* that may be of interest to scholars: notebooks, photographs, drawings, an essay draft of *The Flock* (AU 166), a working draft of *The Flock* (AU 165), a letter explaining Little Pete's love affair with a mixed-blood girl (AU 5171), and more. Austin's lifework also contains many poems, short stories, dramas, and novels featuring sheep, sheepherding, sheepdogs, and male and female shepherds. The illustrations in this afterword are reproduced by permission of The Huntington Library, San Marino, California.

2. Jim was probably Austin's brother.

3. All of the photographs and drawings hereinafter discussed in this afterword come from AU 537.

4. Editors of *Twentieth-Century Literary Criticism* make this claim, plus numerous critics quoted therein. The quote comes from the same collection: Vernon Young, "Mary Austin and the Earth Performance," 1950, 31.

5. Numerous scholars back up Austin's views. The "pristine" Yosemite wilderness had been heavily managed by Native Californians, even before shepherds, but signs of management are not always obvious to the inexperi-

enced (Blackburn and Anderson; Solnit; Nabhan; Snyder 92). Although Solnit credits Indian people with the "park like atmosphere," she never mentions shepherds and calls other early settlers "squatters." Forest managers in British Columbia are now paying sheep producers to control weeds and brush in new tree plantations (Glimp et al.). Reducing trees and brush to grass produces more water, not less, as Muir claimed. Watershed managers in Massachusetts's "Quabbin" are finding that they must cut trees in order to maintain water regeneration and that overpopulations of deer "have been munching" young hardwoods and softwoods, "virtually all of them, down to the ground" (Dizard 11, 25, 36). Suppression of fire in the Sierra has led to "the brushy understory that is so common in the Sierra now" (Snyder 137). In addition, scientists in Great Britain have found sheep grazing beneficial to buttercup and early spider orchid (Frost 1981 and Hutchings 1987 summarized in West 8).

    6. Limbaugh and Lewis, reel 1, letter dated July 19, 1868.

    7. For a more complete analysis of the Muir-Austin conflict over sheep grazing see (Nelson 221–42).

WORKS CITED

Austin, Mary. "AU." See Mary (Hunter) Austin
Collection.
———. *The Children Sing in the Far West.* Boston:
Houghton Mifflin, 1928.
———. *Earth Horizon.* Boston: Houghton, Mifflin, 1932.
———. *The Flock.* Boston: Houghton, Mifflin, 1906.
———. *Isidro.* Boston: Houghton, Mifflin, 1905.
———. *The Land of Little Rain.* 1903. Introduction by
Edward Abbey. New York: Penguin, 1988.
———. *The Land of Little Rain.* 1903. Photographs by
Ansel Adams. Intro. Carl Van Doren. Boston:
Houghton, Mifflin, 1950.
———. "The Last Antelope." *Western Trails: A
Collection of Short Stories.* Ed. and intro. Melody
Graulich. Reno: U of Nevada P, 1987. 202–8.
———. "The Little Coyote." *The Atlantic Monthly* Feb.
1902: 249–54.
———. Mary (Hunter) Austin Collection. The
Huntington Library, San Marino, CA.
———. *One Hundred Miles on Horseback.* 1889. Intro.

by Donald P. Ringler. Los Angeles: Dawson's Book Shop, 1963.

———. "A Shepherd of the Sierras." *The Atlantic Monthly* 86 (1900): 54–58.

———. *Taos Pueblo.* 1930. Photographs by Ansel Adams. Facsimile ed. Boston: New York Graphic Society, 1977.

———. "When I Am Dead." *New Mexico Quarterly* 4 (1934): 234–35.

Bauer, Robert. "Rediscovering Twenty Hill Hollow." Paper presented at the California History Institute: John Muir in Historical Perspective conference. Stockton, CA, April 18–21, 1996.

Blackburn, Thomas C., and Kat Anderson. *Before the Wilderness: Environmental Management by Native Californians.* Menlo Park, CA: Ballena P Publication, 1993.

Bowen, Ezra. *The High Sierra.* The American Wilderness / Time-Life Books. Washington, D.C.: Time, Inc., 1972.

Bredahl, A. Carl, Jr. *New Ground: Western American Narrative and the Literary Canon.* Chapel Hill: U of North Carolina P, 1989.

Dizard, Jan E. *Going Wild: Hunting, Animal Rights, and the Contested Meaning of Nature.* Amherst: U of Massachusetts P, 1994.

Fink, Augusta. *I-Mary: A Biography of Mary Austin.* Tucson: U of Arizona Press, 1983.

Glimp, Hudson A., Donald G. Ely, James Gerrish, Ed Houston, Rodney Knott, Dan Morrical, Bok Sowell, Charles Taylor, and Robert Van Keuren. "Rangelands, Pasture and Forage Crops." *Sheep Production Handbook.* Englewood, CA: American Sheep Industry Association, 1998. 103–28.

Graulich, Melody, and Elizabeth Klimasmith, eds. *Exploring Lost Borders: Critical Essays on Mary Austin.* Reno: U of Nevada, 1999.

Limbaugh, Ronald H., and Kirsten E. Lewis, eds. *The John Muir Papers, 1858–1957.* Microform. Alexandria, VA: Chadwyck-Healey Inc., 1986.

"Mary (Hunter) Austin, 1868–1934." *Twentieth-Century Literary Criticism.* Vol. 25. Detroit: Gale, 1988. 15–46.

Muir, John. "Draft of My First Summer." pre-1911. In *The John Muir Papers, 1858–1957.* Ed. Ronald H. Limbaugh and Kirsten E. Lewis. Microform. Alexandria, VA: Chadwyck-Healey, Inc., 1986. Reel 31.

———. *My First Summer in the Sierra.* 1911. Ed. Edward Hoagland. Intro. Gretel Ehrlich. New York: Penguin, 1987.

———. "Twenty Hill Hollow journal." 1868. In *The*

John Muir Papers, 1858–1957. Ed. Ronald H. Limbaugh and Kirsten E. Lewis. Microform. Alexandria, VA: Chadwyck-Healey, Inc., 1986. Reel 23.

Nabhan, Gary Paul. "Cultural Parallax in Viewing North American Habitats." *Reinventing Nature? Responses to Postmodern Deconstruction.* Eds. Michael E. Soule and Gary Lease, Washington, D.C.: Island P, 1995. 87–101.

Nelson, Barney. "*The Flock:* An Ecocritical Look at Mary Austin's Sheep and John Muir's Hoofed Locusts." *Exploring Lost Borders: Critical Essays on Mary Austin.* Eds. Melody Graulich and Elizabeth Klimasmith. Reno: U of Nevada P, 1999. 221–42.

Norris, Frank. *The Octopus.* 1901. New York: Signet, 1964.

O'Neill, Elizabeth Stone. *Meadow in the Sky: A History of Yosemite's Tuolumne Meadows Region.* Fresno: Panorama West Books, 1983.

Powell, Lawrence Clark. "Mary Austin: *The Land of Little Rain.*" *California Classics: The Creative Literature of the Golden State.* 1971. Santa Barbara: Capra P, 1982. 44–52.

"A Review of *The Flock.*" *The Nation.* 1906. In "Mary (Hunter) Austin, 1868–1934." *Twentieth-Century*

Literary Criticism. Vol. 25. Detroit: Gale, 1988. 17–18.

Rowell, Galen. "Along the High, Wild Sierra: The John Muir Trail." *National Geographic* April 1989: 466–93.

Snyder, Gary. *The Practice of the Wild.* San Francisco: North Point P, 1990.

Solnit, Rebecca. *Savage Dreams: A Journey into the Landscape Wars of the American West.* New York: Vintage, 1994.

Tracy, Henry Chester. *The American Naturists.* New York: Dutton, 1930.

West, Neil. "Biodiversity of Rangelands." *Journal of Range Management* 46 (1993): 2–13.

White, Richard. "'Are You an Environmentalist or Do You Work for a Living?': Work and Nature." *Uncommon Ground: Toward Reinventing Nature.* Ed. William Cronon. New York: Norton, 1995. 171–85.

Wyatt, David. "Mary Austin: Nature and Nurturance." *The Fall into Eden: Landscape and Imagination in California.* New York: Cambridge UP, 1986. 67–95.

Young, Vernon. "Mary Austin and the Earth Performance." 1950. In "Mary (Hunter) Austin, 1868–1934." *Twentieth-Century Literary Criticism.* Vol. 25. Detroit: Gale, 1988. 29–32

Library of Congress Cataloging-in-Publication Data

Austin, Mary Hunter, 1868–1934.

The flock / Mary Austin ;

with an afterword by Barney Nelson.

p. cm.—(Western literature series)

Originally published: Boston : Houghton Mifflin, 1906.

ISBN 978-0-87417-355-0 (alk. paper)

1. Shepherds—California—San Joaquin Valley—History.

2. Ranch life—California—San Joaquin Valley—History.

3. Sheep—California—San Joaquin Valley—History.

4. Austin, Mary Hunter, 1868–1934. I. Title. II. Series.

SF375.4.C2 A88   2001

636.3'083'097948—dc21   2001027394